Cognitive Ergonomics
Contributions from Experimental Psychology

Cognitive Ergonomics
Contributions from Experimental Psychology

Editors

Gerrit C. van der Veer
University of Twente, Enschede, The Netherlands
Free University, Amsterdam, The Netherlands

Sebastiano Bagnara
University of Siena, Siena, Italy

Gerard A.M. Kempen
University of Nijmegen, Nijmegen, The Netherlands

1992

NORTH-HOLLAND
AMSTERDAM · LONDON · NEW YORK · TOKYO

ELSEVIER SCIENCE PUBLISHERS B.V.
Sara Burgerhartstraat 25
P.O. Box 211, 1000 AE Amsterdam, The Netherlands

Library of Congress Cataloging-in-Publication Data

Cognitive ergonomics : contributions from experimental psychology /
  editors, Gerrit C. van der Veer, Sebastiano Bagnara, Gerard A.M.
  Kempen.
      p.   cm.
   Based on papers presented at the Fifth European Conference on
  Cognitive Ergonomics held in Urbino, Italy, in Sept. 1991.
   "Reprinted from the journal, Acta psychologica, vol. 78, nos. 1-3,
  1991"--T.p. verso.
   Includes indexes.
   ISBN 0-444-89504-3
   1. Human-computer interaction. 2. Human engineering.   I. Veer,
  G. C. van der (Gerrit C.) II. Bagnara, Sebastiano. III. Kempen,
  Gerard, 1943-   .
  QA76.9.H85C643   1992
  004.01'9--dc20   91-48454
                                                               CIP

Reprinted from the journal
ACTA PSYCHOLOGICA, Vol. 78, Nos. 1–3, 1991

ISBN: 0444 89504 3

Transferred to digital printing 2005
Printed and bound by Antony Rowe Ltd, Eastbourne

Contents

Text editing

Graphics design

Acta Psychologica 78 (1991) ix
North-Holland

Preface

The history of this volume started in Urbino, Italy, where in September 1991 the Fifth European Conference on Cognitive Ergonomics (ECCE5) was held under the auspices of the European Association of Cognitive Ergonomics (EACE). EACE brings together researchers in the area of Human–Computer Interaction (HCI) from various disciplines, in particular computer scientists and psychologists.

This volume is based on papers selected for presentation at ECCE5, except for the contributions by De Haan et al. and Heydemann et al., which were submitted to *Acta Psychologica*'s Cognitive Ergonomics Section. All papers have been reviewed and revised especially for this volume. The collection illustrates how concepts, theories and techniques from experimental psychology can be applied in the HCI domain.

The first five papers examine various theoretical issues in Cognitive Ergonomics. The ten other papers analyse performance in specific interactive tasks: computer programming and program debugging, database interrogation, text editing, and graphics design.

We are indebted to the following HCI experts for their detailed critical comments on earlier versions of the papers published here: K. Ammersbach, T.R.G. Green, A. Jameson, A.H. Jørgenson, J.-M. Hoc, T. Landauer, F.L. van Nes, L.L. Norros, M. Thüring, and A. Vandierendonck. Finally, we express our gratitude to Elly Lammers for her administrative and editorial support during all stages of the publication process.

Gerrit C. van der Veer
Sebastiano Bagnara
Gerard A.M. Kempen

Theoretical issues

Acta Psychologica 78 (1991) 3–26
North-Holland

Task knowledge structures: Psychological basis and integration into system design *

Hilary Johnson and Peter Johnson

Queen Mary and Westfield College, University of London, London, UK

A major goal of HCI is to assist software designers to construct useful and usable computer systems. One way to approach this problem is to provide software designers with information about the knowledge users utilize in performing tasks. This paper first outlines the psychological basis for our approach to task modelling, Task Knowledge Structures (TKS) and then considers how to achieve the ultimate objective in developing TKS, i.e. to integrate task models into system design by providing information about users and tasks at appropriate stages in design. To facilitate the integration of TA into system design, a survey of designers was conducted to identify where, when and how task data can be represented to aid integration into system design and the designers' requirements for tools to support task model and system design integration. The results from the survey are then summarised and the designers' requirements for TA tools are outlined, followed by a discussion as to how these requirements might be met. Finally, brief details of design specifications for the proposed tools are presented.

Introduction

The main concern of researchers in human–computer interaction (HCI) is to develop models, methods and techniques to assist software designers to construct computer systems which people find useful and usable. One way to approach this goal is to assume that knowing something about how users approach and carry out tasks will aid software designers when making design decisions which will ultimately affect computer system usefulness and usability. As a result task analysis has emerged as an important aid to early design in HCI. It

* We are grateful to ICL URC for funding the project 'The development of task analysis as a design tool', and also to ESPRIT basic research for funding the further development of the psychological basis of TKS.

Requests for reprints should be sent to H. Johnson, Queen Mary and Westfield College, University of London, Mile End Road, London, E1 4NS, UK.

provides an information source from which design decisions can be made and a basis for evaluating designed systems. Task analysis is an empirical method which can produce a complete and explicit model of tasks in a domain, and of how people carry out those tasks. It focuses design on users' tasks and goals, and the methods for achieving those goals. Design focused on these aspects of a user's task domain can result in improved, more usable systems. This is because rather than leaving design for usability to luck and intuition, task analysis methods can be utilized to inform the designer about factors concerning users' tasks that can influence usability in advance of the designers making inappropriate design decisions.

Rosson et al. (1988) have argued that designers consider obtaining information about users and tasks as a major contributor to the generation of design ideas. They distinguish between design idea generation and design development and suggest that design idea generation is facilitated by task/user analysis tools and techniques, while design development is facilitated through the use of specification and prototyping tools. By carrying out task analyses, users and tasks will be taken into account in the formation of the design idea. If this is the case then interface and system design will take place from an informed position on the part of the designer, resulting in the need for less iterations. Our aim is to transform HCI design from trial and error to an informed and principled activity.

Although task analysis is the investigation of what people do when they carry out tasks, an approach to task analysis involves more than simply observing how people perform tasks. It also involves developing a theory of tasks, techniques of data collection, a method of analysing tasks and a representational framework for constructing task models. This paper divides into two halves respectively, reflecting our two main objectives, which are first, to outline in detail the psychological basis for our theory of tasks which underpins both the TKS methodology and the design of the task analysis tools. This is covered in the first part of the paper. The second half of the paper is concerned with our second objective which is to produce tools to support designers undertaking task analyses and consequently integrating task data into system design. Although the composition of the two halves of the paper may appear distinct, it is important to note that the psychological basis for TKS governs both the nature of the data collected and modelled and the specifications for the TA tools.

In the second half of the paper we first briefly summarise a detailed but small-scale survey undertaken to identify both design practices and designers' requirements for task analysis tools. The final part of this section briefly describes some of our initial thoughts about how we might attempt to design the TA toolset and to see if the designers rather general requirements can be met. For a description of the data collection and analysis methods and constructing task models, see Johnson et al. (1988a). Our approach to task analysis known as Task Knowledge Structures (TKS) has been developed from earlier work on Task Analysis for Knowledge Descriptions (TAKD), Johnson et al. (1984). It is to be contrasted with task analysis techniques which are not concerned with knowledge, such as ability profiling (Fleishman and Quaintance 1984), Hierarchical Task Analysis (Annett and Duncan 1967), and other techniques which have an evaluative role in assessing the complexity of task performance but have no explicit method of task analysis. Information about how users plan and carry out goal-directed behaviour is one aspect of users and tasks that can be identified and analysed. Probably the most often cited approach to TA in HCI is the GOMS approach of Card et al. (1983). The underlying philosophy of GOMS is that tasks can be thought of in terms of goal structures, operations, methods and selection rules. Kieras and Polson (1985) extend the GOMS approach by arguing that production rules can be used to model these goals, operations, methods and selection rules. The work of Kieras and Polson (1985), Payne and Green (1986) and Card et al. (1983), are good examples of current evaluative approaches in HCI which incorporate methods of either predicting the time taken to carry out a task or the degree of difficulty of using an interactive computer system. Each of these approaches is capable, in varying degrees, of making recommendations about how proposed system designs can be used in terms of the ease with which users could perform given tasks. All assume some form of task model.

There are two important features to these approaches which distinguish them from TKS; first, they are designed purely as evaluative tools, and consequently are not concerned with design generation. By this we mean that these approaches assume that decisions about what tasks the system should support have been made elsewhere, and also that one or more design solutions have already been proposed. Second, they focus on the evaluation and prediction of user performance and do not detail any particular method of task analysis. In contrast, TKS is concerned to identify the knowledge requirements of tasks and is aimed at assisting in the generation of design solutions, although TKS may also form part of an evaluation methodology.

It should be noted that the range and complexity of tasks with which we have been concerned in the development of TKS have been beyond the level of simple keyboard tasks, neither is the approach restricted to physical tasks. We are concerned with tasks as complex and rich as designing the room layout of houses, producing multi-media documents, collaborative writing and controlling sophisticated building surveillance equipment.

Psychological basis for Task Knowledge Structures (TKS)

In previous papers, most notably in Johnson and Johnson (1987a, b) and Johnson et al. (1988b) we have outlined our theoretical assumptions, derived from and supported by research in cognitive psychology, which underpin Task Knowledge Structures. Three assumptions have been made, concerning treating tasks as concepts to be represented in memory, the structure of entities in tasks and the differential status of those entities in terms of their representativeness or typicality. In this paper we delve a little deeper into the basis and evidence for these assumptions.

To reiterate our first assumption, we believe that TKSs could be represented as concepts, for instance the TKS 'design a house room layout' could be represented as a concept which would be stored in human memory in the same way as are other concepts. The basis for this assumption is that various types of knowledge are represented and structured in memory, for skilled performance (Anderson 1976), motor control (Johnson 1982), language acquisition and comprehension (Karmiloff-Smith 1986a), hierarchical planning (Cromer 1986), problem solving (Karmiloff-Smith 1986b), categorisation (Johnson 1983) and so on, and it seems unlikely that task knowledge is the exception given the goal driven nature of some of the previously referenced work. Additionally, evidence that knowledge about events is represented in memory comes from Schank (1982) who assumes that knowledge of frequently occurring events is structured into meaningful units in memory, as either scripts, or memory organisation packets. Further support for our assumption that tasks are represented as cohesive units in memory comes from the work of Galambos (1986) who conducted a series of experiments which show that people recognise and use structures of events, such as the order, the sequence and the importance of activities within the event sequence to understand, explain and make predictions about these events. Graesser and Clark (1985) also provide support for the notion that task knowledge is represented in memory. They assume that general knowledge structures which represent goals to causal and enabling states, plans for achieving goals, intermediate states and alternative solutions or paths are represented in a conceptual knowledge structure that is used to interpret events. Finally, Barsalou (1982, 1983, 1987) in discussions of the graded structure of categories, (i.e. from good to poor examples) has argued that there are a number of different types of concept, from common taxonomic categories to adhoc and goal-derived categories. In our view the idea of goal-derived categories corresponds with our assumption that there are concepts directly related to tasks which have specific goals that can be utilized in task performance.

Although we might cite the work of Barsalou as an indication that there is some psychological validity for TKSs, we do not share his views about the data-driven nature of categorisation being entirely in the manner of similarity-comparison processes in working memory. Barsalou (1987) puts forward the proposition that conceptual knowledge is not stored in long term memory. There are two implications from this proposition; first, there are no knowledge structures related to concepts which are permanently stored in memory which can be accessed and processed on different occasions. Second, as we have already stated, categorisation is treated in a data-driven manner, not involving long term memory. Therefore, stimuli are interpreted and processed in working memory on each separate occasion, with each occasion being treated as independent. Barsalou's assumption that conceptual knowledge is not stored in long term memory arises from his observation that category typicality and representativeness are of an unstable nature, varying with different contextual information both within and between individuals from one occasion to another. Johnson (1983) has observed a similar phenomenon, and has argued that the context effects arise due to multiple representation of some entities in memory, further exacerbated by different effects arising due to different purposes of carrying out categorisation. This latter point has been eloquently discussed by Medin and Wattenmaker (1987), Lakoff (1987) and McCauley (1987) who from different perspectives all argue for the use of idealized cognitive models and theories of the world as a basis for classification. The main thrust of their arguments is that depending on the models or theory of the world we are following at the time, this will affect how we classify. McCauley in particular is outspoken in his condemnation of views of categorisation, as in Barsalou (1987) and Brooks (1987) which espouse no centralised, or a decentralised view of categorisation. A centralized view of categorisation assumes that there are higher order structures which might guide categorisation (e.g. idealized cognitive models or theories of the world) and thus explain the empirical findings. On the other hand the decentralized view attempts to explain the finding that entities might be judged typical on one occasion and yet not on others, as indications of at best, the instability of conceptual structures and at worst, the possibility of no conceptual structures in long term memory. In terms of task knowledge structures we believe that overriding goals and the tasks that accomplish them will provide one aspect of the user's current model of the world especially since most human behaviour is goal driven. We explicitly assume that TKSs are represented in long-term memory. TKSs represent the different aspects of knowledge that are possessed by a task performer and as such are a convenient summary representation, constructed by the task analyst, to represent the task knowledge held by one or more task performers and which once constructed will provide a basis for task-centred system design.

What are the benefits of assuming that tasks can be represented as concepts? The main benefits are in the extrapolation of the utility of conceptual organisation to task knowledge, i.e. benefits associated with assuming tasks have an internal structure and that task entities differ in how representative they are of the task. These related assumptions are brought into play because we need to be able to describe and predict how the user will carry out tasks and the relationships between the various aspects of task knowledge. The structure in tasks should give us some indication of the task entities which go together (are carried out in parallel), and which prime, precede or follow on from one another. Tasks would be unstructured if the task world formed a set of tasks in which all possible elements and attributes of tasks could co-occur with equal probability combined with all other possible elements or attributes of tasks. This is obviously not the case; task elements or behaviours do not occur independently of one another. Some pairs or even *n*-tuples of task elements are quite probable whereas others are improbable; some groupings of attributes while being logically possible may never occur in reality. Elements of tasks are generally carried out according to some feasible temporal ordering, designated by a plan. An example of feasible temporal ordering might be the case of a builder who is building a house who cannot begin to build until the bricks have arrived. An architect designing the layout of a house cannot design the upstairs layout until s/he knows how many bedrooms are required. The same architect might have to simultaneously consider certain other related task elements. For instance, in designing a bathroom layout, the respective position of the bath is considered at the same time as the positions of the wash-hand basin and toilet. Empirical evidence for the representation of plans in long term memory can be found in the work on programming tasks (Green et al. 1987). Further evidence for structuring in tasks comes from the work of Byrne (1977) on tasks in domains such as cookery and meal planning. Having identified the plan and any alternative plans to be activated depending on the satisfaction of each plan's pre and post-conditions, we can then acquaint the software designer with these task details in order that entities which precede or prime one another can be provided for in an appropriate manner, for instance spatially close on a single screen, or as automatic activation of one after the other, and so on. (For fuller implications of task structure for system design see Johnson et al. 1988b.)

Our view of task structure is similar to that of Garner (1974) in describing how people represent physical objects based on the structure of those objects in the real world. However, the structure of objects in the real world, in terms of their correlated features and the psychological representation of those features has been a matter for debate more recently, see Neisser (1987) for a direct perception view of perception of features. In considering the structure of tasks in terms of correlated features, i.e. in terms of things that go together because they might share some particular features, we must address the wider problem of why some things go together and why some do not. Medin and Wattenmaker (1987) argue that what makes a concept cohesive has received a variety of answers mainly to do with similarity of features. However, they argue that similarity based approaches to conceptual coherence are insufficient to explain the richness of categorisation. They argue for a theory based approach which emphasizes that coherence derives from both internal structure of a conceptual domain and the position of the concept in the complete knowledge base. According to Medin and Wattenmaker concepts are viewed as embedded in theories and are coherent to the extent that they fit people's background knowledge or naive theories of the world. If we view goal-driven behaviour, executed in tasks as one aspect of people's naive theories of the world, then we need to look no further for reasons that task entities go together or are related, than the fact that together they comprise a coherent view of how the task is carried out. More specifically, entities are related through a common goal, irrespective of whether or not they share common perceptual or functional features.

The third assumption which we have made, and which also is a direct implication of treating task knowledge as conceptual knowledge is that some task entities are more representative of the task than are others. The psychological basis for this argument arises from research into categorisation in cognitive psychology in the last fifteen years. In this space of time the traditional view of categorisation has become systematically challenged by both the probabilistic and exemplar views of categorisation. The traditional or classical view dates back to Aristotle and treats categories as having necessary and sufficient criteria which determine whether or not a particular instance is a category member or not. Classification viewed in this light is taken to be all-or-none. 'All' meaning that the instances have all the necessary and sufficient criteria, and 'none' for those instances not having the necessary and sufficient criteria. The probabilistic and exemplar views, challenged the traditional view on the grounds that some category members appeared to be more representative or typical of a category than did others (Rips et al. 1973; Rosch 1973). The empirical evidence for the claim that exemplars differ in typicality comes from Rosch et al. (1976), McCloskey and Glucksberg (1978) and Smith (1978) who argued that grading examples from good to poor examples, is fundamental to predicting how long it takes to classify something as a category member with typical exemplars being identified faster than atypical exemplars. Graded structure is also fundamental to an understanding of the observation that typical exemplars are generated by subjects on request more often than atypical exemplars (Rosch et al. 1976). Typicality also plays a role in learning where typical exemplars are learnt more quickly than atypical instances. For a full review of the three different approaches to categorisation and their theoretical implications, see Smith and Medin (1981) and Johnson (1983).

In terms of TKS we assume that knowledge about objects and their associated actions differ in how representative they are of the TKS of which they are a part. We argue that in the same way that a 'robin' might be a typical instance of the 'bird' category so might 'house plan' be a typical task entity of the task 'design house room layout'. In addition the typical task entities are just those which might constitute central task procedures which are important for successful task performance. Empirical psychological evidence for procedural centrality and action/object representativeness in task behaviour has been obtained by Leddo and Abelson (1986) who found that for tasks such as borrowing a book from a library there were particular task segments or procedures which were more central to, and more representative of, going to the library than other segments.

Knowledge in TKSs

A TKS is a summary representation of the different types of knowledge that are recruited and used in task behaviour. A TKS is related to other TKSs by a number of possible relations. For a full discussion of the knowledge in TKSs, see Johnson et al. (1988b).

One form of relation between TKSs (within role relations) is in terms of their association with a given role. A role is assumed to be defined by the particular set of tasks that an individual is responsible for performing as part of their duty in a particular social context (a 'job' is one form of social context). A person may take on a number of roles, for example 'author', 'referee', 'teacher' etc. There are tasks associated with each of these roles. There will be a TKS for each task that a role may be required to perform. Those tasks that are related because they are performed by the same role will have the role relation property associating their respective TKSs.

A second form of relation between TKSs (between role relations) is in terms of the similarity between tasks across roles. This occurs when a person is performing a similar task under different roles. For example, a person may carry out the task of 'arranging meeting(s)' while assuming the role of 'chairman', 'client', 'husband'. Each task may be performed differently in one or other respect. However, a single person assuming all these roles would have a knowledge structure for each task and also knowledge (not necessarily explicit) of the relations between these tasks. Thus the TKSs for each of the 'meeting arranging' tasks would have relational links to other 'meeting arranging' tasks performed by that person when assuming a different role.

Within each TKS there are other knowledge representations. The goal structure identifies the goals and subgoals contained within the TKS. A goal is assumed to be the desired state of affairs that a particular task can produce and forms part of a person's conceptualisation of their task world while carrying out a given role. A subgoal is a lower level goal. The goal structure also includes the enabling and conditional states that must prevail if a goal or subgoal is to be achieved. In this way the goal structure might be thought of as a kind of plan for carrying out the complete task: It is the particular formulation and possible ordering of subtasks that are undertaken to achieve a particular goal. The goal structure is carried out through a procedural structure which may contain alternative procedures for achieving a particular subgoal. A procedure is an executable unit of behaviour that forms part of a subtask. It contains control and execution statements such as iteration, sequence and branching. As there may be alternative sets of procedures for achieving a particular subgoal there are also conditional and contextual groupings of these procedures. In this way strategies are represented. Strategies are particular sequences of procedures. The procedures rely on other knowledge of the objects and actions which when combined constitute a given procedure statement. Actions and objects which are characterized by sets of attibutes, are the lowest level of elements of tasks.

Action and object knowledge is represented in a taxonomic substructure. Within the taxonomic substructure information about the properties of the action(s) and object(s) are represented. This includes, the representativeness of the object, the class membership and other attributes such as the procedures in which it is commonly used, its relation to other objects and actions, its features (such as what the object can do or possess) and its typicality rating. These are described in more detail in Johnson et al. (1988a) along with a number of techniques for identifying and modelling TKS components. Johnson et al. (1988b) provide a method for applying these task models to design using examples from real tasks that have been analysed and real designs in complex domains that have been proposed.

Superficially, TKS might appear to have certain similarities to GOMS in particular in respect to the goal structure. However, there are fundamental theoretical and practical differences. To reiterate an earlier point, TKS is distinct from GOMS in that GOMS can be construed as an evaluative method, not pertinent for design generation. TKS is appropriate for design generation because it encompasses techniques for collecting, analysing and modelling the task knowledge that people possess. This information will then provide a set of requirements which can be taken into account in devising a design solution by indicating which parts of tasks need support. GOMS does not, by contrast, have a methodology for identifying task knowledge, rather it models the knowledge required to use a particular design solution. The theoretical underpinning to TKS is also very different from that of GOMS. We make explicit claims about treating tasks as conceptual structures in memory. Additionally, GOMS makes no assumptions about the structure of tasks, nor does the GOMS approach have any views about object representativeness or procedural centrality. Following the GOMS approach, there is no reason to assume that any particular part of a task is any more central or important to the successful execution of the task than any other, whereas there is empirical evidence, cited above, that such notions do have validity.

Finally, with GOMS usability is related to how long it takes for an expert to complete a task, where elements of that task performance are considered to be additive and there is a component for 'thinking time'. GOMS, as extended by Kieras and Polson (1985) argues that complexity can be measured by counting production rules. However, different classes of rules are not considered. Usability in TKS relates to system support for representative and central task elements, where usability is deemed to decrease depending on the level of support for task procedures. Empirical evidence for the improvement in usability afforded by modelling TKSs is provided by Davis (1988) in a small-scale pilot study of graph and table drawing. The first part of the study identified representative objects and actions, central procedures and sequencing of task procedures for the above two tasks across a population of 12 subjects. An experiment was then carried out in which three further groups of subjects were required to undertake graph and table drawing tasks using one of three different interfaces with the same underlying functions.

One interface was structured so that it positively supported representative and central task elements, and task sequencing identified by the TKS modelling stage. The other interface was unstructured; representative and central task elements were supported but representative objects were not identified with their associated actions and the sequencing of task procedures was not supported. A further control group had an interface which contained neither central nor representative task features, and which had no explicit task structure.

The results of this pilot study showed that subjects found the structured interface easier to use, and also this interface had a higher preference value from subjects. Additionally, the structured interface resulted in quicker task execution, and the resulting graph and table drawings were better quality in that they were more complete. Also the unstructured but central and representative interface design produced better performance than the control group interface, but less so than the structured interface group. The findings support our theoretical view that TKSs provide important information about users that can be used to design improved user interfaces.

Task analysis and system design

In the first half of the paper we outlined the psychological basis of TKS, contrasted TKS with other approaches to task analysis and discussed the knowledge which was assumed to comprise TKSs. The theoretical assumptions made in TKS have implications for the composition of the data collection and analysis techniques we have devised (Johnson and Johnson 1987a, b). These techniques therefore support system designers in identifying task structure and representative and central task elements, and also in constructing task models (TKSs).

In this section of the paper we consider how TKS might contribute to system design. In order to facilitate the introduction and use of task analysis in system design we need to know what software design practice entails, and also how to provide tools to support designers in carrying out task analyses. A current view of design processes and practices in system design is provided by Johnson and Johnson (1989) who carried out in-depth interviews with three system designers who were considered to be representative of designers by their respective companies. Two purposes of the interviews are of relevance here; first, this provides a view of what these designers considered to be the potential contribution of task analysis to various stages of system design and thereby, an idealized view of how task analyses might eventually contribute to system design. This idealized view of the potential contribution of task analysis to system design is being taken account of in our extended version of TKS (Johnson and Johnson 1991a). The second purpose of the interviews, as an initial stage in our intention to produce on-line task analysis tools, was to identify designers' tool requirements. Below we summarise some of the survey findings. (Full details of the survey can be obtained from Johnson and Johnson (1989).)

The designers who participated in the survey thought that the contribution of TA might be:

At the feasibility / initial planning stage:
- Identifying and documenting any new functions/new tasks the computer may support.
- Identify potential functionality of the system from user perspective.
- Identify user population and characteristics of that population.
- Identify characteristics of interface to be developed.
- Allocation of function between user and system.
- Assess scope/degree of larger-scale TA to be undertaken later in development lifecycle.

At the requirement / analysis stage:
- Identify and document user/UI requirements comprising details about:
- hierarchical structure of tasks (goals and subgoals);
- how users achieve goals and subgoals;
- listing and ordering of undertaking task procedures;
- frequency with which particular procedures were carried out by users;
- reasons why and circumstances under which one procedure was used in preference to another;
- inputs and outputs from each procedure;
- events, data used, actions, objects;
- standard set of properties relating to objects and actions, e.g. frequency, time taken, etc.;
- expectations the user entertains about the system after user has carried out an operation;
- division between user and system.

At the Design stage / User interface Development / Dialogue design:
- Provide initial input to guide dialogue and screen design, comprising:

- details of what users expect to have available to them at any one time;
- the structure and sequence of their usage of system facilities;
- the names and form of representation to be given to screen-presented objects and events;
- information that should be available in given contexts, (i.e. design of screens);
- structure between contexts, (i.e. mapping between screens);
- how much to put on the screen at once with reference to number of commands;
- what information should go on screens and the grouping of that information;
- what commands are needed to support user operations and what those commands will be;
- user testing.

At the prototyping stage:
- Guide initial format and presentation of prototype by indicating how the screens should look.
- Identify data that has to be displayed.
- Identify operations and sequencing of procedures.
- Ensure dialogue specification is represented in a format that can be understood and verified with end users and to carry this out.

At the validation stage:
- User testing.

At the update and maintenance stage:
- Identifying, documenting and cataloguing user problems.

This is an idealized view of the potential contribution of task analysis to system design, and could be taken as a specific set of requirements of an approach to task analysis. The designers were also asked about the desired characteristics and criteria they thought that the *proposed* TA tools must satisfy; the results are summarised below:

- easy to use,
- on line,
- can be used on a small-scale at first then used more extensively later,
- must have a facility for registering and defining objects and actions,
- facility to list characteristics and attributes of objects,

- facility to help in constructing procedures and task hierarchy,
- facility to show task hierarchy and procedures graphically – and allow procedures to be expanded into detailed operations by some diagrammatic convention,
- assist in process of transforming task model into representations used by designers.

Developing a task analysis toolset

Knowledge Analysis of Tasks (KAT) is the name we have given to our framework for task analysis and the resulting toolset. The focus of the KAT tools are the data collection and analysis techniques, and the designers' tool requirements. The designers' tool requirements however are very general and it was not easy for the designers to say exactly what they needed. In addition, a particular problem for the designers was in stating how the task data could be represented in an optimal manner such that integration to system design is possible (see Johnson and Johnson 1989). Because the designers' tool requirements were very general, we have in the design of the specificiation for the tools, mainly taken account of what aspects of a TKS it is important for designers to gather data about and which aspects are repetitive and time consuming, such as generating generic (common across occurrences) task elements. However, it was of benefit to identify what tool requirements designers had, especially as this is seldom done, and thus provide for an informed and principled approach to the development of the tools. We will consider later whether any of these general requirements are met.

It is important to stress that the tools are at present under construction. There is a wide range of techniques that can be employed (Muller 1991) to ensure that the tools once implemented will be useful and usable. Not only must the tools be easy to use, but their deployment should also result in better designed systems, therefore empirical studies to this end are also planned. (A brief outline of the tool design specifications is given later; for a fuller account, see Johnson and Johnson 1990.)

With respect to satisfying the designers' general tool requirements, the specifications we have put together are for on-line tools thus satisfying one of the designers' requirements. Another requirement was that the tools must be easy for designers to use. As previously stated, the resulting tools will be tested extensively in order to assess how easy they were to use. The TA tool specifications incorporate introductory sections to task analysis to aid the system designers. The tools will support the construction of a task model which can be undertaken to a lesser or greater extent thereby making the tool appropriate for small scale use at first and more extensive use later. An object/action database or dictionary will allow TA tool users to register, define and characterize attributes of objects and actions and their interrelationships. A goal-oriented structure to be completed by designers should, with the addition of appropriate help facilities, aid in the construction and graphical representation of task goal structures and their relation to specific procedures.

Further requirements for the task analysis tools, resulting from the survey were that the proposed tools must be versatile and flexible, capable of integration with different approaches to, and stages in design, should not impede design stages and should be capable of integration into both structured design methods and other design lifecycles. These requirements stem from our finding that design practices vary considerably from designer to designer, even within the same design team, with some stages being eliminated, grouped together and so on. The content and function of the task analysis tools should provide end-user task information and users' interface requirements, relate identified requirements back to end-users in a non-technical fashion, and lessen the need for designers to have to rely only on their intuition. The TA tools should also provide the basis for getting user feedback, decrease the number of end-user problems associated with using the system in undertaking tasks and provide a basis for including end-users in quality assurance procedures.

A further requirement, in order to aid integration into system design methods is that an independent subset of the proposed tools should be specifically tailored to one of several potential structured design methods. SSADM was chosen as a target structured design method for our tool subset. The reason why SSADM was chosen is that it is a UK government standard, and our collaborators at the time, ICL, were in the process of training their designers in the use of SSADM. Therefore, there was an opportunity to introduce designers to our tools at the same time. Several approaches could be taken to relating task analysis to system design. It might, for instance, be advantageous to represent task knowledge in one or more of the data representation formats of SSADM and other structured design methodologies, i.e. data flow diagrams, entity life histories and so on. However, when we asked designers if this was appropriate, each of them indicated that the task data should be kept separate. The approach which was taken by us was to identify for all phases of SSADM whether task data was indeed appropriate, what that data would consist of, how it would be collected and modelled and what *specific* stage of SSADM it would be address. Therefore, we identified for the phases of SSADM the potential contribution of task data. In the present version of our tool specifications, we have already identified for all the stages in the feasibility, system analysis and system design phases of SSADM where task analysis would be appropriate and what that contribution would be.

The KAT toolset

In this paper we are only able to briefly outline some of the features which characterize the proposed tools. In the full report (Johnson and Johnson 1990) we indicate how the tool screens might look, the menu labels and content, and also describe the functions that will be available to the users, and/or will be carried out automatically by the implemented TA tools. Here we only have space to provide brief details of the screen contents and functions.

The format of the TA tool screens will be as follows: First, screen titles, centred at the top of the screens will identify and refer to each of the individual screens. Shaded menu headers situated below the screen titles, will be visible at all times when the screen has been selected. Each of the menu headers will comprise pop-up menus with menu items which will be activated when the menu header is selected. Therefore all the menu items will be available to the user at any time that the user chooses, on selection of the appropriate menu header. Some of the menus will also have within them cascading menus. On the user selecting a menu item the pop-up menu will disappear and the appropriate text as a result of the menu selection will appear on the screen. Text on the screen will be presented as on-line scrollable text. Below we give a general indication of the contents of each of the screens.

A total of four screens and a number of associated windows will complete the KAT toolset. SCREEN 1 KNOWLEDGE ANALYSIS OF TASKS (KAT) will always be present when the application is started. This first screen will be mainly an information screen giving general details about KAT and the KAT tool. This screen will provide an introduction to KAT activities. When the user actually wants to undertake these activities, other screens will be activated automatically as a result of the user selecting either 'Colecting data', 'Analysing data' or 'Modelling tasks'. In addition, screen 1 provides the user with a short-cut to the TKS screen (see later). Screen 1 will comprise six menu headers with pop-up menus within these headings. The menu headers will be *'Information'*, which will provide details about reports and papers which give details about the theoretical underpinning to KAT and other sources of information. The second header, *'Task analysis'* will have five items on a pop-up menu providing (a) a general introduction to task analysis, (b) general guidelines to follow when carrying out task analyses, (c) a glossary of task analysis terms, (d) details about the scope of a task analysis and finally, (e) a list of the knowledge components which comprise TKS.

The third header of screen 1 will be *'Data Collection'*, which will comprise two items on a pop-up menu, (a) an introduction to data collection and (b) a menu item 'Collecting data' which when selected will advise the user that the 'Data collection and analysis screen' will be opened. The fourth header of the KAT screen will be *'Data Analysis'* again with two menu items, (a) an introduction to data analysis, and (b) a menu item 'Analysing data' which again will activate the 'Data collection and analysis screen'. The fifth menu header *'Task modelling'* comprises three menu items,(a) an introduction to task modelling, (b) an automatic activation of the TKS screen, labelled 'Modelling tasks', and (c) automatic activation of the system integration screen, labelled 'System integration'. The final header of screen 1 will be *'TKS'* which will have four menu items; (a) 'Create)', (b) 'Edit', (c) 'Delete', and (d) 'TKS'. Choosing any of these menu items will provide automatic activation of the TKS screen. On choosing 'Create' the computer once on the TKS screen will ask for the title and also about the goal and taxonomic substructures, which are substructures of the TKS (see Johnson and Johnson, (1987b) for a discussion of what these relate to). Choosing 'Edit' or 'Delete', results in the user being asked for the title of the TKS to be edited/deleted which is then displayed on the screen. Choosing 'TKS' will activate the screen without any further activity until the user pulls down and selects an item from a menu on that screen.

THE DATA COLLECTION AND ANALYSIS SCREEN comprises four menu headers, *'Data collection', 'Knowledge sources', 'Collection techniques'* and *'Data Analysis'*. The pop-up menu for the Data collection' header has four items, (a) an introduction to data collection, (b) an introduction to knowledge components which make up TKS, (c) a menu item 'Actions and Objects' which also has a cascading menu to items giving information about how to identify actions and objects and their representativeness and typicality, and to accessing the 'Object database' and the 'Action database'. On the user selecting the 'Object database' or the 'Action database' either the Object database window or the Action database window respectively, will appear on the screen. For detailed descriptions of the functions of these windows we would refer readers to Johnson and Johnson (1990). Briefly the windows provide a register of the actions and objects and their characteristics which can be added to, amended, queried and so on. The final item on this menu is (d) containing details about procedures and plans such as their identification and their relationship to goals and subgoals.

The second menu header *'Knowledge sources'*, has two menu items, (a) details of background literature on how to find out about people's tasks, and (b) a menu item 'People' which has a cascading menu to details about choosing people to participate in a task analysis in terms of their number, expertise, characteristics and also about eliminating subject bias. The third menu header *'Collection techniques'* has eight menu items covering (a) introduction to collection techniques, (b) interviews with cascading menu to information about interviews, their advantages, disadvantages, appropriateness and also a proforma questionnaire. The same information is able under separate menu items for (c) observation (d) concurrent protocol analysis (e) retrospective protocols, (f) the use of concurrent protocols versus retrospective protocols, (g) the problems with protocol analysis and finally, (h) problems with collection techniques generally.

The final menu header of the 'DATA COLLECTION AND ANALYSIS SCREEN' is *'Data analysis'* which has two menu items (a) an introduction to data analysis, (b) details about generification with a cascading menu to introducing generification and the identification of generic actions and objects. The goal structure item on this cascading menu will activate the goal structure window which will support the inputting of plans, procedures, goals and subgoals, which it will then present to the user in both graphical and textual (including rule-based) form. A final item on this menu is computer-supported generification. Computer supported generification procedures will be supported by a generification window which when activated will identify generic actions and objects from those entered by the user, dependent on a user-defined threshold. As a result the window will display a comprehensive list of generic objects and actions, listed alphabetically and with the corresponding frequency also quoted. This list can then be transferred to the taxonomic substructure.

THE TASK KNOWLEDGE STRUCTURES (TKS) SCREEN will display the empty slots of the TKS which have to be completed by the designer. Extensive help facilities will be available at this screen to indicate how the slots can be filled and referring to each of the other screens and windows for techniques and methods by which the TKS might be completed. There are five menu headers, *'TKS', 'Roles', 'Tasks', 'Taxonomic substructure'* and *'Goal substructure'*. Under the *'TKS'* menu header are five items, (a) an introduction to TKS, (b) a dictionary consisting of a scrollable list of already constructed TKSs, (c) a menu item Object database which on selection will activate the 'Object database window', (d) a menu item Action database which on selection will activate the 'Action database window', (e) a menu item System integration which will on selection automatically activate the System integration screen. Under the 'Roles' menu header there are three menu items allowing the user to (a) create, (b) edit and (c) delete role relationships. The *'Tasks'* menu header similarly has three menu items allowing the user to (a) create, (b) edit and (c) delete links between TKSs. The *'Taxonomic substructure'* and *'Goal substructure'* menu headers have menu items which introduce the user to the substructures and allow them to be created, edited and deleted and for the automatic transference of items within the two databases to the goal and taxonomic substructures. Two windows are associated with each of these menu headers (for full details please see the report referred to earlier).

Finally, once the designer is satisfied with the partially complete or fully completed TKS, the task model can then be transformed into one of the various representational formats which will make up the System integration screen. The SYSTEM INTEGRATION TOOLSET SCREEN has three menu headers, *'Information', 'Task modelling representational formats'* and *'SSADM'*. Under the *'Information'*, menu header there are two menu items, (a) which discusses the potential integration of task analysis and (b) KAT, to the feasibility/initial planning, requirement/

analysis, design, prototyping, validation and update and maintenance stages of a design lifecycle. The *'Task modelling representational formats'* menu header will have four menu items, (a) an introduction to the different formats TKS could be represented in, (b) a view of the use of English as a medium for TKS representation, (c) the use of the diagrammatic/graphical representation of TKS given by the goal substructure window, and finally (d) the use of production rules which can be constructed directly from the TKS via the goal substrucutre window. The last menu header of the SYSTEM INTEGRATION SCREEN *'SSADM'*, has four menu items which cover (a) introduction to SSADM, the contribution of KAT at the feasibility stage (b), at the systems analysis stage (c) and systems design stage (d).

This completes the brief details of the proposed KAT screens and windows. We have taken into account the aspects of TKS which need to be identified, analysed and modelled according to the theoretical underpinnings of TKS. We have also attempted to meet the very general requirements requested by the designers in our survey data. In the next stage of the research the KAT tool designs will be tested by using a variety of techniques to see how well they satisfy designers' requirements. Folowing satisfactory testing the tools will then be implemented.

Summary

At the beginning of this paper we argued that one way of assisting designers to build usable and useful computer systems was to provide them with information about how users plan and carry out goal directed behaviour. We then discussed the uses of task analysis in system design, followed by a description of the theoretical background to TKS. The results of a survey concerned with identifying what designers wanted, needed and expected in the way of tools to support task analyses and their integration into system design, were presented. However the requirements were very general and the survey small-scale so it will be necessary to extensively test the paper based specifications before implementation. Finally, we provided a very brief description of a set of design specifications for proposed KAT tools to support designers in carrying out task analyses and their integration into system design. Other research being conducted on TKS involves the development of a formal notation to express TKS, incorporation into an advanced prototyping environment and also the use of formal methods including algebraic expressions, first and second order logic to express TKS elements. In addition to this research in a new project concerning interactive dialogues for explanation and learning, the framework of TKS is being extended to provide users of information systems with satisfactory explanations which allow them to successfully use those systems and learn from the interaction.

References

Anderson, J.R., 1976. Skill acquisition. Cambridge, MA: Harvard University Press.

Annett, J. and K.D. Duncan, 1967. Task analysis and training design. Journal of Occupational Psychology 41, 211–221.

Barsalou, L.W., 1982. Context-independent and context-dependent information in concepts. Memory and Cognition 10, 82–93.

Barsalou, L.W., 1983. Ad hoc categories. Memory and Cognition, 11, 211–227.

Barsalou, L.W., 1987. 'The instability of graded structure'. In: U. Neisser (ed)., Concepts and conceptual development: Ecological and intellectual factors in categorization. Cambridge: Cambridge University Press.

Brooks, L.E., 1987. 'Decentralized control of categorization; the role of prior processing episodes'. In: U. Neisser (ed)., Concepts and conceptual development: Ecological and intellectual factors in categorization. Cambridge: Cambridge University Press.

Byrne, R., 1977. Planning meals: Problem solving on a real data-base. Cognition 5, 287–332.

Card, S.K., T.P. Moran and A. Newell. 1983. The psychology of human–computer interaction. Hillsdale, NJ: Erlbaum.

Cromer, R., 1986. Hierarchical planning disability in the drawings and constructions of a special group of severely aphasic children. Personal communication.

Davis, S., 1988. Knowledge structures in the human computer interface. Unpublished manuscript. Queen Mary College, University of London.

Fleishman, F. and G. Quaintance, 1984. Taxonomies of human performance. New York: Academic Press.

Galambos, J.A., 1986. 'Knowledge structures for common activities'. In: J.A. Galambos, R.P. Abelson and J.B. Black (eds.), Knowledge structures. Hillsdale, NJ: LEA.

Garner, W.R., 1974. The processing of information and structure. New York: Wiley.

Graesser, A.C. and L.F. Clark, 1985. Structures and procedures of implicit knowledge. Norwood, NJ: Ablex.

Green, T.R.G., R.K.E. Bellamy and J.M. Parker, 1987. 'Parsing and gnisrap: A model of device use'. In: H.J. Bullinger and B. Shackel (eds.), INTERACT '87. Proceedings of the Second IFIP Conference on Human–Computer Interaction. Amsterdam: North-Holland.

Johnson, H., 1983. Categorisation in children and adults. Unpublished Ph.D. thesis, Birmingham University.

Johnson, H. and P. Johnson, 1987a. The development of task analysis as a design tool: Collecting and analysing task data. Report to ICL.

Johnson, H. and P. Johnson, 1987b. The development of task analysis as a design tool: A representational framework for task modelling. Report to ICL.

Johnson, H. and P. Johnson, 1989. Integrating task analysis into system design: Surveying designers' needs. Ergonomics, 32, 1451–1467.

Johnson, H. and P. Johnson, 1990. The development of task analysis as a design tool. Design specifications for tools to support task analyses. Report to ICL, January.

Johnson, H. and P. Johnson, 1991a. Extending TKS for system design. Manuscript in preparation.

Johnson, P., 1982. Functional equivalence of images and movements. Quarterly Journal of Experimental Psychology 34A, 349–365.

Johnson, P., 1985. 'Towards a task model of messaging'. In: P. Johnson and S. Cook (eds.), People and computers; Designing the user interface: Cambridge: Cambridge University Press.

Johnson, P., D. Diaper and J. Long, 1984. 'Tasks, skill and knowledge; Task analysis for knowledge based descriptions'. In: B. Shackel (ed.), Interact '84. Amsterdam: North-Holland.

Johnson, P. and H. Johnson, 1991b. 'Knowledge analysis of tasks: Task analysis and specification for human–computer systems'. In: A. Downton (ed.), Engineering the human–computer interface. London: McGraw Hill.

Johnson, P., H. Johnson and F. Russell, 1988a. Collecting and generalising knowledge descriptions from task analysis data. ICL Technical Journal 6, 137–155.

Johnson, P., H. Johnson, R. Waddington and A. Shouls, 1988b. 'Task related knowledge structures: Analysis, modelling and application'. In: D.M. Jones and R. Winder (eds.), People and computers: From research to Implementation. Cambridge: Cambridge University Press, pp. 35–62.

Karmiloff-Smith, A., 1986a. 'Some recent issues in the study of language acquisition'. In: J. Lyons, R. Coates, M. Deuchar and G. Gazdar, New horizons in linguistics, 2. Harmondsworth: Penguin.

Karmiloff-Smith, A., 1986b. 'Children's problem solving'. In: A. Brown and M. Lamb (eds.), Advances in developmental psychology, Vol III. Hillsdale, NJ: Erlbaum.

Kieras, D. and P.G. Polson, 1985. An approach to the formal analysis of user complexity. International Journal of Man–Machine Studies 22,365–394.

Lakoff, G., 1987. 'Cognitive models and prototype theory'. In: U. Neisser (ed)., Concepts and conceptual development: Ecological and intellectual factors in categorization. Cambridge: Cambridge University Press.

Leddo J. and R.P. Abelson, 1986. 'The nature of explanations'. In: J.A. Galambos, R.P. Abelson and J.B. Black (eds.), Knowledge structures. Hillsdale, NJ: LEA.

McCauley, R., 1987. 'The role of theories in a role of concepts'. In: U. Neisser (ed.), Concepts and conceptual development: Ecological and intellectual factors in categorization. Cambridge: Cambridge University Press.

McCloskey, M. and S. Glucksberg, 1978. Natural categories: Well-defined or fuzzy sets? Memory and Cognition 6, 462–472.

Medin, D. and W.D. Wattenmaker, 1987. 'Category cohesiveness, theories and cognitive archeology'. In: U. Neisser (ed.), Concepts and conceptual development: Ecological and intellectual factors in categorization. Cambridge: Cambridge University Press.

Muller, M., 1991. PICTIVE: An exploration in participatory design. Human Factors in Computing Systems (CHI '91), 225–231.

Neisser, U., 1987. 'From direct perception to conceptual structure'. In: U. Neisser (ed.), Concepts and conceptual development: Ecological and intellectual factors in categorization. Cambridge: Cambridge University Press.

Payne, S.J. and T.R.G. Green, 1986. Task-action grammars: A model of the mental representation of task languages. Human–Computer Interaction 2, 93–133.

Rips, L.J., E.J. Shoben and E.E. Smith, 1973. Semantic distance and the verification of semantic relations. Journal of Verbal Learning and Verbal Behavior 14, 665–681.

Rosch, E., 1973. 'On the internal structure of perceptual and semantic categories'. In: T.E. Moore (ed.), Cognitive development and the acquisition of language. New York: Academic Press.

Rosch, E., 1978. 'Principles of categorisation'. In: E. Rosch and B. Lloyd (eds.), Cognition and categorisation. Hillsdale, NJ: Erlbaum.

Rosch, E., 1985. 'Prototype classification and logical classification: The two systems'. In: E.K. Scholnick (ed.), New trends in conceptual representation: Challenges to Piaget's theory? Hillsdale, NJ: Erlbaum.

Rosch, E., C. Mervis, W. Gray, D. Johnson and P. Boyes-Braem, 1976. Basic objects in natural categories. Cognitive Psychology 8, 382–439.

Rosson, M.B., S. Maass and W.A. Kellogg, 1988. The designer as user: Building requirements for design tools from design practice. Communications of the ACM 31, 1288–1298.

Schank, R.C., 1982. Dynamic memory: A theory of reminding and learning in computers and people. New York: Cambridge University Press.

Smith, E.E., 1978. 'Theories of semantic memory'. In: W.K. Estes (ed.), Handbook of learning and cognitive processes, Vol 6. Potomac, MD: Erlbaum.

Smith, E.E. and D. Medin, 1981. Categories and concepts. Cambridge, MA: Harvard University Press.

Acta Psychologica 78 (1991) 27–67
North-Holland

Formal modelling techniques in human–computer interaction *

G. de Haan [a], G.C van der Veer [a,b] and J.C. van Vliet [a]

[a] *Free University, Amsterdam, The Netherlands*
[b] *Twente University, Enschede, The Netherlands*

This paper is a theoretical contribution, elaborating the concept of models as used in Cognitive Ergonomics. A number of formal modelling techniques in human–computer interaction will be reviewed and discussed. The analysis focusses on different related concepts of formal modelling techniques in human–computer interaction. The label 'model' is used in various ways to represent the knowledge users need to operate interactive computer systems, to represent user-relevant aspects in the design of interactive systems, and to refer to methods that generate evaluative and predictive statements about usability aspects of such systems. The reasons underlying the use of formal models will be discussed. A review is presented of the most important modelling approaches, which include External–Internal Task Mapping Analysis; Action Language; Task-Action Grammar; the Goals, Operators, Methods and Selection model; Command Language Grammar and Extended Task-Action Grammar. The problems associated with applying the present formal modelling techniques are reviewed, and possibilities to solve these problems are presented. Finally, we conclude with a discussion of the future work that needs to be done, i.e., the development of a general design approach for usable systems, and the need to focus attention on the practice of applying formal modelling techniques in design.

Introduction

The use of computers by non computer experts has sharply increased during the last decades. For this group of users, the computer is only a tool to get their work done, comparable to a pencil or a notebook. As such, to learn and use the computer system is an additional task derived from the application of this specific tool in the course of performing the primary task: the work being done (Van der Veer 1990), and therefore the effort of having to learn and use a computer system should be minimal. In Cognitive Ergonomics (the study of human–

---

* Requests for reprints should be sent to G.C. van der Veer, Dept. of Computer Science, Free University, De Boelelaan 1081a, 1081 HV Amsterdam, The Netherlands.

computer interaction) the subject of study is the usability of computer systems to perform the user's tasks. From the viewpoint of the user, operating a computer system is not merely pressing buttons, but rather building an understanding of the system, as if it were a human partner (Oberquelle 1984). To analyse the computer system in this sense, a number of concepts have been introduced. Oberquelle (1984) has introduced the virtual machine, as the functionality of a system in terms of abstract (hence: virtual) functional units and their behaviour, without considering the details of the implementation and the hardware. In a similar way, the term user interface is used to include all perceptually and conceptually relevant elements and behaviour of a computer system that the user might know about, and should know to perform his tasks successfully. One might say that ideal or competent users, who have full knowledge of how to use a computer system to accomplish all possible tasks, know everything about the user interface in the Cognitive Ergonomic sense. In the remainder of our contribution we will use the term computer system with the same meaning as, and interchangeable with, the terms 'virtual machine' and user interface, to denote the system as far as it is relevant for a user, and hence may be either perceived directly, or may be conceptually conceived by interacting with the system.

In cognitive ergonomics, the usability of a system is assumed to depend on the organizational circumstances in which the computer is employed, on characteristics of the intended users of the computer system, on the tasks they have to perform, on the style of the dialogue with the computer system, and on the physical environment. Matters of concern are, among others, job and task analysis, task and computer experience, skill and problem solving, and measurements of office equipment. In Cognitive Ergonomics it is generally agreed upon that any system designed to be used by people should meet certain requirements. For example, Gould and Lewis (1985) state that a computer system should be:

- functional,
- easy to use,
- learnable,
- pleasant to use.

A computer system can be said to be usable to the extent that it meets these requirements. Usable systems, then, should provide the users with the functions they need to fullfil their tasks (functionality). The operation of the computer system should not require extensive mental or physical effort (easy to use). The operating procedures of the system should be easy to learn and easy to remember after periods of not using the system. And finally, using the computer system should be enjoyed by users. The joy of using computers will not be dealt with in this paper, because it is difficult, if at all possible, to approach this question with formal techniques.

It is necessary to distinguish the general concept of usability of computer systems in the sense of fulfilling the first three (or even all four) of the above-mentioned requirements, and the separate requirements. We will reserve 'usability' and 'usable', to refer to the quality of the system as a whole, and reserve 'easy to use' and 'ease of use', to refer to the narrow sense of demanding little mental or physical effort.

Formal modelling techniques or approaches can be used to represent the knowledge the user needs and/or the actions the user should perform to delegate his tasks to the computer system. We will not distinguish between the terms technique and approach and use these terms interchangeably. What is important is that applying formal modelling techniques results in models of knowledge and behaviour, which can be analyzed to investigate the extent computer systems fulfil the three requirements for usable systems.

Models and levels of abstraction in human–computer interaction

In order to develop a better understanding of what is involved in designing usable systems, it is necessary to take a closer look at the role of the user in operating a complicated device such as a computer, and introduce the notion of a user's mental model. Norman (1983) distinguished between three types of models: the user's mental model, the system image and the conceptual model of a computer system.

(a) The user's mental model is a model of the machine that users create or, according to Norman, which naturally evolves when learning and using a computer. This type of model does not have to be, and usually is not, accurate in technical terms. Instead, it may contain mis-conceptualizations, omissions and it does not have to be stable over time. However, a user's mental model is indispensable for the user to plan and execute interaction with the system, to predict, evaluate, and explain the behaviour of the computer, and to reduce the mental effort involved.

(b) The user's mental model is based on the system image, which includes all the elements of the computer system the user comes into contact with. As such, the system image includes all the aspects ranging from the physical outlook of the computer and connected devices to the style of interaction and the form and content of the information exchange. Although Norman (1983) excludes teaching materials and manuals from the definition of the system image, these could be included as well because they also shape the image of the system.

(c) The third type of model that Norman distinguishes is the conceptual model of the target system. This is the technically accurate model of the computer system created by designers, teachers and researchers for their specific purposes. As such, this type of model is an accurate, consistent and complete representation of the system, as far as user-relevant characteristics are involved.

The important point of Norman's distinction between three types of models is that in well-designed systems the conceptual model of the designers forms the basis for the system image, which in turn is the basis for the user's mental model. A good design starts with a conceptual model derived from an analysis of the intended users and the users' tasks. The conceptual model should result in a system image and training materials which are consistent with the conceptual model. This, in turn, should be designed in order to induce adequate users' mental models.

Formal modelling techniques in human–computer interaction are used to represent the knowledge users need about the operation of a proposed computer system, and to describe the actions users have to perform to delegate tasks in order to attain their task goals. By analyzing these representations, something can be said about whether the design of a computer system meets the required functionality, and to what extent the design will be easy to use and learnable. Formal representations show the complexity of the knowledge a user needs to acquire in relation to the tasks to be delegated.

Users need knowledge about a computer system for 'translation' in two directions. First, users come to the computer system with a set of tasks, and they will have to know how to translate and rephrase task delegation into the operating procedures and commands provided by the computer system. Secondly, after a command has been supplied to the system, the user must know how to interpret the behaviour of the system, and how to determine the success or failure of the task attempts. That is, users have to know how to translate their highly abstract task goals into the physical actions towards the computer system, and know how to relate the physical responses of the system to task goals and to task knowledge. In human–computer interaction it is common to distinguish several levels of abstraction in this specification/interpretation cycle (e.g., Moran 1981; Nielsen 1986; Frohlich and Luff 1989). Nielsen proposed the following levels of abstraction in the knowledge of computer users:

- Task Level,
- Goal Level,
- Semantic Level,
- Syntax Level,
- Lexical Level,
- Physical Level.

The essence of the notion of levels is that a user and a computer system communicate via certain types of languages that are different from 'natural' language. The correspondence between the meaning in the user's mind and the physical exchange of tokens between user and machine is far from trivial.

For example, a secretary who receives the instruction (task) to make a copy of a letter sent to a customer must know that copying a letter means to reproduce it on paper (goal level), which, in terms of a computer system, means to send a particular text file to a printer device (semantic level). It has to be known that commands must be submitted to the computer by specifying the operation (e.g. printing), a delimiter, an ordered list of arguments (e.g., printer destination first, letter identification second) and an end-of-command indicator (syntax level). Further, this user has to know that the letter is called 'smith.txt', that the command to send something to a printer is called 'print', and the name of the particular printer (lexical level). Finally, to submit the command the appropriate keys have to pressed (physical level).

Especially where it concerns larger and more complicated computer systems, it will be more difficult for a user to have a complete and flawless knowledge of each of these levels. Only expert users will have a mental model which will be consistent with the conceptual model of the system. In terms of Norman (1983), the user's mental model will usually differ from the conceptual model of the system. However, when a user interface is consistently structured and allows for a clear and straightforward mapping between the levels of abstraction then it will be easier for a user to develop an adequate mental model of the system. The key question is then: how to design user interfaces in such a way that the development of an adequate user's mental model is stimulated.

Methods to enhance the usability of computer systems

Once a system is designed and implemented, this topic may be approached relatively straightforward. When a computer system is successfully used, one may assume that the system did indeed stimulate the development of adequate mental models. For design purposes, empirical testing of the final product alone is not very practical; empirical testing is difficult and costly, it requires extensive effort, and its results come very late in the development cycle, often even too late to influence the design (Reisner 1983). A number of methods and tools has been developed to enable predictions regarding the end product's usability at an early stage of the design process.

One approach is to use design methodologies which involve active user cooperation during the whole process of design, to avoid that the design differs very much from what it was meant to be. Proposals for design methodologies like this frequently involve iterations to account for changing system requirements which may arise when users get familiar with using the new system.

Rapid prototyping can also be used to determine if a system, or parts of it, are usable. A prototype simulates the behaviour of the proposed machine more or less accurately (Diaper 1989).

Another approach is to use analogies and metaphors, which means that the design is built around an image borrowed from everyday life, like the furniture and tools in a clerk's office (e.g. the desk-top metaphor). Using a familiar image aims at letting users build on established knowledge, without having to start from scratch to build a mental representation of a system.

A technique that designers expect to be really practical for improving the quality of design is to use guidelines and standards, or simply to look at the products of successful competitors. Applying guidelines like 'never use more than six colours' may save a large amount of time and effort otherwise spent on user testing (see for example Smith and Mosier 1984).

Finally, one may use modelling techniques to capture and analyze the knowledge the user needs to delegate tasks on a proposed computer system. Using these techniques, the analyst makes a more or less formal representation of the relations between the task-goals of the user and the operations needed to reach these goals. By analytic methods that estimate the complexity and consistency of the formal model, e.g. counting the number of rules or calculating the average number of parameters of each rule, something can be said about the functionality, ease of use and learnability of the proposed system.

Above, the various methods to ensure and improve the quality of designs have been presented separately. In actual design, however, combinations of methods need to be used, because no method can resolve all design questions, whereas combinations might produce better results. Which particular method to use depends on the specific question and the ability of a method to solve it, and on how much time and effort is needed to apply the method (Preece and Keller 1990).

In the following, we will concentrate on the use of formal modelling techniques to model user knowledge. We expect that formal modelling techniques have a number of advantages the other methods do not have, even though their use is not as well established as some other methods. The main advantage of formal models is the possibility to specify a design very precisely, without ambiguity. Using a formal notation also creates the possibility to automate parts of the implementation of user interfaces, and provides more possibilities for rapid prototyping and user testing. Moreover, analytic methods applied to formal models (aiming at establishing indexes of ease of use, learnability, and functionality) do not require a working system, as opposed to using empirical measurement techniques. Because only an initial specification of the user interface is required, formal modelling techniques may be applied very early in design to predict some of the usability aspects of the system. Also, since neither users nor working systems are involved, formal modelling can be applied at relatively low cost in comparison with techniques that require more than just a specification.

We do not mean to say that empirical testing is not needed at all. Rather, we would like to suggest that applying formal models could enable the answer to certain questions in design earlier and with less effort than empirical testing, thereby leaving more resources for other methods to improve the quality of the design.

A further advantage, which formal techniques share with using analogies and metaphors, is that they closely adapt to Norman's notion of mental models. Formal models and metaphors both describe aspects of the conceptual model of the system and aspects of the user's mental model, but metaphors give an informal account of the conceptual model, and formal models represent aspects of it (e.g., the 'how-to-do-it' knowledge of a competent user) formally. Whereas metaphors refer to the conceptual meaningfulness of mental models, formal representations present the structure and consistency of an adequate mental model (i.e., a mental representation that is compatible with the conceptual model). The other advantages of formal models combined with their close connection with the notion of users' mental models makes them suitable candidates to serve as conceptual models to base design upon.

An overview of formal modelling techniques

In this overview, the most important or well-known of the formal modelling techniques to represent user knowledge will be treated. Applying formal models for user interface design includes some variant of a general procedure:

(a) the analyst makes a list containing what the users' goals can be and what users have to do in order to reach these goals;
(b) a model is built from the goal-operation sequences, using a more or less formal representation language, e.g., Backus-Naur Form;
(c) the model is restructured to represent the knowledge of a target group of users, such as 'novices', 'competent users', 'occasional users'. Many techniques model the knowledge of the 'ideal user', who is assumed to have full knowledge of operating the system and does not make any errors;
(d) the model is analyzed using some metric, e.g., providing indications of complexity of the transcription rule system, of the discriminabil-

ity of sets of tasks or sets of objects, and of the consistency of representations at different levels of abstraction.

The differences between the various kinds of models include characteristics of the formal language (conceptual basis for representation), the levels of abstraction used in representing the communication between the user and the computer, and the methods used for the analysis of ease of use, learnability, and functionality. The most important reason behind these differences relates to the main goal the model is constructed for. Various categorizations of formal modelling approaches have been proposed (e.g. Green et al. 1988; Murray 1988; Oberquelle 1984; Rohr and Tauber 1984; Simon 1988; and Whitefield 1987).

The main problem in comparing these different categorizations of approaches to formal modelling is that they distinguish the modelling techniques to a large extent on scientific dimensions derived from the field of research, instead of practical considerations. For example, Green et al. (1988) ask whether a model describes performance or competence aspects of behaviour. Murray (1988) distinguishes prescriptive and descriptive models. In a similar vein, Simon (1988) uses the degree of idealization to distinguish between modelling ideal behaviour and real behaviour. Finally, Nielsen (1990) and Whitefield (1987) distinguish models, based on whether the model is owned or created by the user, the designer, the computer system, or the researcher, and what or who is being modelled.

Our intention is to analyze modelling approaches in relation to their merits for design from the point of view of usability. As systems are designed in order to enable task delegation by users to systems, we will base our categorization on this phenomenon. In delegating tasks to a computer four aspects are of importance.

(a) *External tasks*: Users have to perform tasks existing in a task domain outside the computer, which have to be rephrased in terms of the tasks that can be delegated to the computer.

(b) *User knowledge*: In delegating tasks, users need knowledge about the computer system, about the objects and operations the system knows about, and how to operate these in terms of physical actions.

(c) *User performance*: User performance is concerned with the users' behaviour in delegating tasks to the computer system. Users must perform certain actions, both mental, perceptual and physical.

(d) *The computer system*: The system is the actual tool for task delegation, and, as a side effect, a main source for the user's knowledge of
the interaction. It is the goal of the design process.

In accordance with these four aspects of task delegation, formal
modelling techniques will be divided into four categories, each with its
own specific purpose: task environment analysis, user knowledge analysis, user performance prediction, and representation for design purposes. The assignment of models to particular categories is not mutually exclusive. A formal representation can (and sometimes will) be
used for different purposes, but assigning models to the category for
which they were primarily developed will enable a fair judgement of
their advantages as well as their restrictions.

Models for task environment analysis

With task environment analysis we apply a modelling technique that
focusses on the characteristics of how to execute tasks in a certain task
domain, and related knowledge of this task domain. The single example
we show in this category is 'External–Internal Task Mapping' (ETIT,
Moran 1983).

ETIT

Moran's External–Internal Task Mapping analysis is meant to
analyze the relations between the external task domain (which refers to
the tasks a user sets himself, or are set for a user, in relation to
everyday reality) and the internal task domain (representing the delegation of suitable tasks to a computer system designed for application in
the external task domain). Fig. 1 contains an example of text manipulation as a user's task, related to text manipulation using a simple editor
on a computer (for a description of the editor, see Moran 1983).

The example shows the entities and the operations or tasks involved
in the two contexts. In the external task space, several object types are
referred to, such as characters, lines, and tasks are known, such as
adding, moving, removing. The example editor, however, only knows a
single object type 'string' as an entity that can be 'inserted', 'cut', or
'pasted'. In the analysis, several task–object combinations are excluded, because they do not make sense, such as an operation to split
characters. Exempting these irrelevant operations, a number of map-

EXTERNAL TASK SPACE

Terms: Character, Word, Sentence, Line, Paragraph (Text)
Tasks: Add, Remove, Transpose, Move, Copy, Split, Join

Excluded: Copy, Split and Join Characters

INTERNAL TASK SPACE

Terms: String
Tasks: Insert, Cut, Paste

MAPPING RULES

1. Split, Join Sentence → Change Text
2. Text   → String
3. Add   → Insert
4. Remove   → Cut
5. Transpose   → Move
6. Split   → Insert
7. Join   → Cut
8. Change Text   → Cut String + Insert String
9. Move Text   → Cut String + Paste String
10. Copy Text   → Cut String + Paste String + Paste String

Fig. 1. External–Internal Task Mapping analysis.

ping rules can be determined which state how to translate a particular task from one environment to the other. The task 'copy a sentence' in the external world can be mapped on a task delegated to the editor, 'copy a string', which in its turn, must be rewritten as a combination of the actions 'cutting a string', 'inserting the string back in its original location', and 'inserting it elsewhere'. According to Moran, establishing the mapping between the objects and operations of the external and the internal tasks will make it possible to make inferences about the functionality, learnability and consistency of the user interface. ETIT should also be applicable in assessing the extent in which transfer of knowledge will occur between different user interfaces. Although ETIT is mentioned in the literature many times, we presently do not know whether ETIT has ever been applied to real systems.

Models to analyze user knowledge

The modelling techniques in this category employ a formal grammar to analyze and represent the knowledge the user needs to operate a user interface. This type of model may be used to compare the usability of different interfaces or different design options, and to predict differences in learnability.

More specifically, these techniques describe and analyze the knowledge the user must have in order to translate his tasks (originally represented by a user at the semantic or conceptual level) into the appropriate physical actions required to operate the system. We mention two modelling techniques in this category, Reisner's Action Language (Reisner 1981, 1983, 1984) and Task-Action Grammar (TAG, Payne 1984; Payne and Green 1986). Both techniques use a formal grammar to describe the task-action mappings, and both assess usability aspects by counting the number of rules, the depth of the derivation of rules and the number of exceptional rules. They differ with regard to the formal grammar they use.

Action language

Reisner's Action Language represents the task-action mappings in a notation called Backus-Naur Form (BNF), named after two of its authors (Backus et al. 1964). BNF is a formal notation to describe phrase structure grammars by means of a number of hierarchically organised rules. BNF is well known in computer science, where it is used, among other purposes, to describe what the legal or grammatically correct expressions are in programming languages like Algol and Pascal. In BNF, each rule specifies the relation between the more abstract term on the left-hand side and the more specific terms on the right-hand side by means of the 'is-defined-as' operator ($::=$). Alternatives are indicated by the 'or' symbol (|). In various extensions of the notation, succession may be indicated by the 'sequence' symbol (+), and options are enclosed in brackets ([...]). BNF is a notation for context-free grammars which means that terms on the left-hand side are uniformly rewritten on the right-hand side independent of other terms and rules. As such it is not possible to indicate that, for instance, the form of a verb in English depends on whether the subject of the sentence is singular or plural.

In Cognitive Ergonomics BNF can be used to describe the legal sentences in the communication language the user has to use to delegate tasks to the computer system. In this way it models what a user has to know. Reisner (1983) extended BNF to include cognitive actions, written in angle brackets ($\langle\ \rangle$), and physically observable

employ Dn	:: =	⟨retrieve info. on Dn syntax⟩ + use Dn
⟨retrieve info. on Dn syntax⟩	:: =	⟨retrieve from human memory⟩ | ⟨retrieve from external source⟩
⟨retrieve from human memory⟩	:: =	⟨RETRIEVE FROM LONG-TERM MEMORY⟩ | ⟨RETRIEVE FROM SHORT-TERM MEMORY⟩ | ⟨USE MUSCLE MEMORY⟩
retrieve from external source	:: =	RETRIEVE FROM BOOK | ASK SOMEONE | EXPERIMENT | USE ON-LINE HELP
use Dn	:: =	identify first line + enter Dn command + PRESS ENTER
identify first line	:: =	...
enter Dn command	:: =	TYPE D + type n

...

Fig. 2. Action Language (Reisner 1983).

actions, written in capital characters. Fig. 2 shows a fragment of Reisner's Action Language or psychological BNF.

The first line in the example shows that the issuing of the command 'Dn' (to delete n lines) consists of a cognitive action (to retrieve the correct syntax of the command), followed by a plain non-terminal (referring to how the syntax information is used and which keystrokes are involved). Although retrieving the needed information is a cognitive activity, and using it a as physical activity, both parts are rewritten in the same way. Reisner's action language has not been extensively used; Richards et al. (1986) used it to specify a graphical operating system shell (MINICON) with it. The only other application we know of stems from Reisner's own work on the pre-cognitive action language (Reisner 1981), in which two version of a drawing program are compared, one that does, and one that does not treat all the data objects in a uniform way. In this study, the non-uniform interface was characterized by the presence of additional rules to describe the exceptions, and as it was predicted, this interface turned out to be more difficult to learn and use than the interface which needed less rules to be completely described.

Reisner's work has indicated that BNF can be used to describe the knowledge the user needs to operate a computer system. However, in terms of the strength of expression, more powerful grammars can be, and are, used. Shneiderman (1982) introduced the idea of using a 'multi-party BNF' for representing the interaction decomposition regarding both 'partners' in human–computer interaction (see Innocent et al. (1988) for an elaboration of this concept). These formalisms, again, have not yet been elaborated for real-life situations. But BNF-like grammars of this type are still restricted to representation of sets of single rules.

A further development of BNF, Van Wijngaarden grammars (Van Wijngaarden et al. 1969), provides a formal representation technique for structured grammars that include the use of two levels of production rules. Payne and Greene (1983) show that set grammars (related to Van Wijngaarden's two-level grammars) enable the representation of 'family-resemblances' among rules. Only this new type of representation could account for the perception of consistency and inconsistency in syntax constructions, and, hence, could be used as a better model of a user's perception of an interaction language. This analysis led to the development of TAG.

TAG

Task-Action Grammar (Payne 1984; Payne and Green 1986) employs a more sophisticated semantic feature grammar than Reisner's BNF. 'Simple tasks' in TAG are represented by rules which can be rewritten in the same way as BNF rules, but in addition, TAG contains features which make it possible to describe tasks in terms of the meaning they may have for the user. In technical terms this means that it is possible to have rules describing the structure of sets of rules, which is not possible in the original versions of BNF. In terms of the user this means that tasks such as 'moving the cursor to the left' or 'moving the cursor to the right' are identical except for the indication of the direction. In fig. 3 cursor movement is used to illustrate Task-Action Grammar. The commands are listed in the 'Dictionary of simple tasks', from which a simple 'Rule schema' is derived that

List of commands

move cursor one character forward	ctrl-C
move cursor one character backward	meta-C
move cursor one word forward	ctrl-W
move cursor one word backward	meta-W

List of features	*Possible Values*
Direction	forward, backward
Unit	character, word

TAG definition

Dictionary of simple tasks
  move cursor one character forward
      {Direction = forward, Unit = char}
  move cursor one character backward
      {Direction = backward, Unit = char}
  move cursor one word forward
      {Direction = forward, Unit = word}
  move cursor one word backward
      {Direction = backward, Unit = word}

Rule Schemas
  Task [Direction, Unit] → symbol [Direction] + letter [Unit]
  symbol [Direction = forward] → 'ctrl'
  symbol [Direction = backward] → 'meta'
  letter [Unit = word] → 'W'
  letter [Unit = character] → 'C'

Fig. 3. Task-Action Grammar (Green et al. 1988).

illustrated the consistency of the syntax of the example. The user needs only knowledge of one general rule and of the 'features' Direction and Unit.

Green et al. (1988) have applied Task-Action Grammar to describe and explain the results of various experiments on command languages. An experiment is reported, in which subjects had to learn and use three applications, which were supplied with similar and different command languages. Learnability predictions were established for various formal modelling techniques, and for several design guidelines. A comparison between the predicted and the actual results showed TAG's predictions to be most accurate. Finally, Green and co-workers have applied TAG to describe several commercially available software packages, from which the general conclusion is drawn that extensions are needed when TAG is used for other purposes than the analysis of command language consistency.

Models of user performance

Methods for user performance predictions are the modelling techniques primarily targeted at analyzing, describing, and predicting user behaviour and time needed to get tasks done while using a particular computer system. Two often cited modelling approaches in this category are the GOMS model (Goals, Operators, Methods and Selection Rules) of Card et al. (1983) and the Cognitive Complexity Theory (CCT) of Kieras and Polson (1985).

Internally, the models used in this category are not very different from the models used to analyze the knowledge of the user, except for the fact that these models have a formal production-rule (if... then...) representation instead of a formal grammar. They do, however, differ with respect to the purpose of application. Whereas modelling techniques to analyze user knowledge describe and analyze what a user should or must know (without specifying how a user should apply this knowledge), the techniques to predict user performance describe and analyze what a user should know and, additionally, what a user should actually do in order to attain task goals.

As such, the GOMS and the CCT models are performance models, whereas the Action Language and the TAG representation are competence models. This difference may be illustrated for the case when the user may choose from alternative methods. In Reisner's Action Language and TAG the choice from alternatives is just described and specified by the 'or' (|) symbol, as a complete list of different possibilities, without indicating any conditions for actual choices. In GOMS and CCT, however, the goal is to predict user performance, and consequently the conditions for a user to choose an option must be specified in advance, e.g., by inferring individual users' strategies from observation.

The most serious implication from this is that GOMS and CCT require a complete specification of the task goal hierarchy of the user. Another consequence of this choice is, that GOMS and CCT implicitly claim that they can formally represent much more than Action Language and TAG claim, namely actual behaviour, instead of only

```
GOAL: Edit-Manuscript
. GOAL: Edit-Unit-Task-until no more unit tasks
. . GOAL: Acquire-Unit-Task
. . . Get-Next-Page-if at end of page
. . . Get-Next-Task
. . GOAL: Execute-Unit-Task
. . . GOAL: Locate-Line
. . . . [select: Use-String-Search-Method
. . . .           Use-Linefeed-Method]
. . . GOAL: Modify-Text
. . . . [select: Use-Delete-Word-1-Method
. . . .           Use-Delete-Word-2-Method
. . . .              . . . ]
. . . . Verify-Edit
```

Fig. 4. GOMS top level of an editing task (Card et al. 1983).

knowledge as a basis of behaviour. The task goal hierarchy which is needed for a GOMS or CCT analysis is in both cases a GOMS representation, or a hierarchical specification of the users' goals, operators, methods and selection rules.

GOMS

Fig. 4 presents an example of a GOMS representation of part of a text editing task.

As can be seen in the example, goals exist at several different levels of a task. A 'general goal' like editing a manuscript is initially subdivided into 'unit task goals', which correspond to the tasks the user knows how to perform. In general, 'unit tasks' correspond to the commands of a computer system. such as deleting a word, transposing two words, etc. in case of an editor. Unit task goals are further subdivided into a number of levels of 'subgoals', until they can be resolved by applying 'operators' or the 'elementary perceptual, motor, or cognitive acts', such as pressing a key, inspecting the screen or acquiring the next unit task. As a matter of fact, GOMS forms a family of models, because the level at which operators are defined may vary, and this level defines the granularity of a GOMS model.

Methods like using a string search or repeatedly pressing 'linefeed' to get the cursor in position are collections of operators. If there is more than one method to reach a goal, then selection rules determine which method will be used. For example, if the target position of the cursor is on the screen, the linefeed method is used, otherwise, the string search method is used. The time predictions GOMS generates depend on the level at which the operators are defined. In general, the predictions are based on the summation of the times needed to execute the elementary actions of the model, which include physical acts (pressing a key), perceptual acts (locating the cursor), and cognitive acts (making a selection).

A well-known member of the GOMS family of models is the Keystroke Level model. This model, however, lacks the analysis purpose of GOMS itself and is purely meant for predicting error-free, expert performance times. In the Keystroke Level model the user's tasks are analyzed at the level of unit tasks. The time to perform each unit task is estimated by adding the time to acquire the unit tasks, the time to execute the keystrokes in the associated commands, and the time needed for mental operators, which are inserted into the command sequences according to sets of heuristic rules. Fig. 5 shows an example of keystroke level analysis. General time parameters are estimated for different actions like 'press key or button', 'point with mouse', and 'mentally prepare'. The unit task illustrated requires the performance of a sequence of actions ('reach for mouse', 'point to word', 'select word' etc.), for which the corresponding time parameters are added to estimate the total execution time.

GOMS is probably the most cited formal model in human–computer interaction, even though GOMS is meant to be applied under rather restrictive conditions and for a rather limited purpose.

(a) As a performance model GOMS is restricted to predicting error free performance. In GOMS the cognitive load of a user is assessed by counting the number of active goals in memory. Lerch et al. (1989) used GOMS-based estimates of mental overload to predict error behaviour, which they showed to be valid for simple ('overload') errors, although GOMS still is unable to predict conceptual errors. This is a serious restriction. Roberts and Moran (1984) report that experts spend between 4 and 22 per cent of their time in correcting (only) serious errors (non-experts would struggle with errors even more often). Although Card et al. (1983) mention that it should be possible to apply GOMS to include error repair and non-expert performance, this has, to our knowledge, not been investigated in their studies.

Operator	Description and remarks	Time (sec)
K	PRESS KEY OR BUTTON.	
	Pressing the SHIFT or CONTROL key counts as a separate K operation. Time varies with the typing skill of the user; the following shows the range of typical values:	
	Best typist (135 wpm)	0.08
	Good typist (90 wpm)	0.12
	Average skilled typist (55 wpm)	0.20
	Average non-secretary typist (40 wpm)	0.28
	Typing random letters	0.50
	Typing complex codes	0.75
	Worst typist (unfamiliar with keyboard)	1.20
P	POINT WITH MOUSE TO TARGET ON A DISPLAY.	1.10
	The time to point varies with distance and target size according to Fitts's Law, ranging from 0.8 to 1.5 sec, with 1.1 being an average. This operator does *not* include the (0.2 sec) button press that often follows. Mouse pointing time is also a good estimate for other efficient analogue pointing devices, such as joysticks.	
H	HOME HAND(S) ON KEYBOARD ON OTHER DEVICE.	0.40
$D(n_D, l_D)$	DRAW n_D STRAIGHT-LINE SEGMENTS OF TOTAL LENGTH l_D CM.	$0.9n_D + 0.16l_D$
	This is a very restricted operator; it assumed that drawing is done with the mouse on a system that constrains all lines to fall on a square 0.56 cm grid. Users vary in their drawing skill; the time given is an average value.	
M	MENTALLY PREPARE.	1.35
$R(t)$	RESPONSE BY SYSTEM.	t
	Different commands require different response times. The response time is counted only if it causes the user to wait.	

Method for Task T1-BRAVO:

Reach for mouse	H[mouse]
Point to word	P[word]
Select word	K[YELLOW]
Home on keyboard	H[keyboard]
Issue Replace command	MK[R]
Type new 5-letter word	5K[word]
Terminate type-in	MK[ESC]

$$T_{execute} = 2t_M + 8t_K = 2t_H + t_P = 6.2 \text{ sec.}$$

Fig. 5. Keystroke model (Card et al. 1983).

(b) The model generates reasonably good predictions under rather specific conditions only. Card et al. (1983) base their predictions on GOMS analyses adapted to individual subjects, for instance, to account for differences in the criteria of selection rules. In the validation studies of the GOMS model, the subjects had to make changes to manuscripts from annotations, which is a rather limited task domain. Lerch et al. (1989) applied their GOMS analysis on a restricted set of tasks (financial calculations) for two commercially available spreadsheet systems.

(c) The GOMS analysis depends very much on the definition of unit tasks, but 'task the user knows how to perform' is not a precise definition, so that the analysist may have to rely on his own intuition in dividing the task into unit tasks (Wilson et al. 1988). In order to resolve this problem, Van der Veer (1990) defines unit tasks as 'elementary primary task that may not be decomposed into other primary tasks', where 'primary tasks', in turn, are defined as 'task the user wishes to perform, independent of the specific characteristics of the tools he will use'.

Although there are serious criticisms about how GOMS is applied to predict user behaviour, it is one of the most widely investigated models, and the value of GOMS as a heuristic method to gain insight in the users' tasks must not he underestimated. A GOMS representation is a very useful and systematic tool to describe the structure of decomposing the user's tasks in smaller elements. For structural description, a GOMS representation is much more useful than models of user knowledge or models for task environment analysis.

CCT

Another model to predict user performance is Cognitive Complexity Theory from Kieras and Polson (1985). CCT is primarily an implementation of the GOMS model in terms of an explicit production system; that is, Kieras (1988) has published a set of rules to rewrite the implicit if-goal then-action notation of a GOMS model into a real production system notation. Such a system is called a 'job-task analysis' which is regarded as a description of the process going on in the users' working memory. In this representation multiple conditions can be taken together by the AND operator, while the action part may consist of more than one action, including cognitive actions, such as the creation of (new) subgoals. An example of a fragment of a CCT analysis is shown in fig. 6, which represents a particular way to delete words, when using IBM's Displaywriter (Kieras and Polson 1985: pp. 374). The first

Method to delete a single word

```
(PDELW1
IF (AND        (TEST-GOAL delete word)
               (NOT (TEST-GOAL move cursor to %UT-HP %UT-VP))
               (NOT (TEST-CURSOR %UT-HP %UT-VP)))
THEN (         (ADD-GOAL move cursor to %UT-HP %UT-VP)))

(PDELW2
IF (AND        (TEST-GOAL delete word)
               (TEST-CURSOR %UT-HP %UT-VP))
THEN (         (DO-KEYSTROKE DEL)
               (DO-KEYSTROKE SPACE)
               (DO-KEYSTROKE ENTER)
               (WAIT)
               (DELETE-GOAL delete word)
               (UNBIND %UT-HP %UT-VP)))
```

Fig. 6. Cognitive Complexity Theory (Kieras and Polson 1985).

production shows that the condition (goal is to delete a word, the goal position of cursor movement is not identical with the current position of the cursor) leads to the addition of the subgoal to move the cursor to the goal position. The second production shows that the condition (goal is to delete a word, cursor is at the goal position) leads to the sequence of actual actions that imply deletion and to the removal of the goal once this is fulfilled.

The aim of this representation is threefold:

(a) By estimating the time needed to execute productions, and, in particular, the operator parts of it, time predictions can be generated, based on the actual tasks and the task-job representation.
(b) Analyzing what the user has to do to operate a certain system in terms of production rules provides a uniform way to compare computer systems. When two computer systems have comparable functionality, the system that requires most production rules will be more difficult to learn and use. Kieras and Polson have extended this point by stating that the ease of learning a new system will depend on the number of common rules between the new and the known system. Transfer of learning would only depend on the number of common rules, irrespective of, for example, confusion created by seeming commonality.

(c) The production system representation can be used to analyze task-to-device mapping, representing the difference between the task-goal hierarchy of the user (how to do it) and the state transition of the system (how it works). Although, presumably no one will deny that performance is best when the user's expectations coincide with the behaviour of the system, Kieras and Polson have treated this subject too scant to say anything conclusive about it.

CCT may be used to compare systems in more or less the same way as the GOMS model and the models to analyze user knowledge or task environment. Because CCT uses the GOMS model, the same criticisms apply: the model has difficulty to cope with errors (Vossen et al. 1987), so that it may only be applied to routine expert tasks. According to Knowles (1988), CCT, by virtue of its reliance on GOMS is restricted in application to tasks involving no problem solving, besides that CCT relies on the quantitative aspects of representing knowledge at the expense of qualitative aspects. At present, CCT has been used mainly to analyze transfer of training effects both successfuly (Polson and Kieras 1985; Foltz et al. 1988; Polson 1988) and less successfuly (Vossen et al. 1987).

Critics state that CCT is not very clear about what actually constitutes a single production: 'Production rules can be rewritten in many different forms, thereby affecting the apparent complexity in terms of number of rules, number of times each one is used, etc.' (Green et al. 1988).

Models of the user interface

Models in this category are developed in relation to formal techniques for design specification. The models in this group aim at providing a complete and full representation of the 'virtual machine' (the computer system as seen from the point of view of a fully competent user). These models represent aspects of a computer system that are relevant for both the potential user and the designer, at the different levels of abstraction of human–computer interaction.

This category is exemplified by the Command Language Grammar (CLG; Moran 1981) and by Extended Task-Action Grammar (ETAG; Tauber 1988, 1990). Both are methods to describe the hierarchical structure of a user interface at the levels of abstraction mentioned before, in a related way. Apart from details such as naming, ETAG and CLG differ with respect to the formalism they use, and in some choices related to the organization and the main points of the representation.

CLG

Starting with a formal description of the tasks and the associated task-entities, and finally ending with the specification of the physical actions, the user interface can be precisely described for design purposes in a top-down manner. Moran (1981) partitions the communication between man and machine into three components, each containing two levels. Each of the six levels is a complete description of the computer system at its level of abstraction:

Conceptual Component:	Task Level
	Semantic Level
Communication Component:	Syntactic Level
	Interaction Level
Physical Component:	Spatial Layout Level
	Device Level.

Moran (1981) only discusses the first four levels and leaves the other two for future elaboration (and to 'classical' Ergonomics). The division into three components clearly indicates the major concern at each pair of levels (see Van der Veer 1989), but this is not strictly needed to understand CLG, and it will not be discussed here.

The division into six levels is directly related to considerations of good user interface design, based on the user's mental model. That is, the user comes to the computer system to get tasks done, and in order to do that, the tasks of the user have to be rephrased into the task language of the computer system and finally specified by physical actions of the user. The other way around, in order to understand the system, the user has to perceive the physical signals of the system, code them into meaningful symbol structures, and rephrase the responses of the system in terms of his primary tasks. Each of Moran's levels describes at a particular level of abstraction an aspect of this process, in terms of what has to be translated and how it is to be done. In this way, the output of each level is a further refinement of a previous level, or, in the opposite direction, an abstraction of the next level.

The purpose of the representation at the task level is to analyze the user's needs and to structure the task domain in such a way, that a computer system can play a part in it. The task level describes the structure of the tasks which can be delegated to the computer system. In order to use an interactive system, the user has to translate his tasks into operations the computer knows about.

The representation at the semantic level describes the set of objects, attributes, and operations, the system and the user can communicate about for the purpose of task delegation: for the system as data structures and procedures, and for the user as conceptual entities and operations on them.

The syntactic level describes which conceptual entities and operations may be referred to in a particular command context or system state, and how that is done, in terms of linguistic aspects (references to commands and objects, including the lexicon) and lexicographic aspects (the order of referencing, display areas). At this level it is specified, for instance, that there is a window to position delegation commands, and that the command to delete is 'delete' followed by the type of arguments to delete and a list of arguments.

Ultimately, the communication between man and computer is a matter of physical actions, such as sequences of key presses, movements of the mouse, meaningful signals like the 'beep' etc. The interaction level describes the translation of the reference names of commands and objects into the associated physical actions and the structure of the interaction, including typing rules and mouse manipulation conventions and the reactions and prompts from the system.

In Moran (1981), this is also the level where the treatment of CLG ends, but he adds that a full CLG analysis would also include a specification at the spatial layout level, and one at the device level. The former level describes the arrangements of the input and output devices, including display graphics, while the device level would describe all the remaining physical features. Fig. 7 presents a fragment of an electronic mail tool, described at the four highest levels of interaction. At the task level a description is shown of the task to read new messages (if any) and of the objects SEND-MESSAGE (which indicates a new message) and MESSAGE (which indicates any message, whether old or new). One of the constituents of the above mentioned task (check for new mail) is subsequently represented at semantic, syntax, and key-stroke level.

According to Moran (1981), three different points of view apply to CLG.

```
NEW-MAIL =(A TASK (* Check for new SEND-MESSAGEs,
               if any, read them)
            DO (SEQ: (CHECK-FOR-NEW-MAIL)
                     (READ-NEW-MAIL)))

SEND-MESSAGE = (AN ENTITY
               NAME ='Send-message'
               (* comments...))

MESSAGE =  (AN ENTITY
            REPRESENTS (A SEND MESSAGE)
            NAME ='Message'
            AGE = (ONE-OF: OLD NEW)
            (* comments...))

SEM-M2 =   (A SEMANTIC-METHOD
            FOR CHECK-FOR-NEW-MAIL
            DO (SEQ:  START EG-SYSTEM)
                     (SHOW DIRECTORY)
                     (LOOK AT DIRECTORY FOR
                     (A MESSAGE AGE = NEW))))

SYN-M2 =   (A SYNTACTIC-METHOD
            FOR CHECK-FOR-NEW-MAIL
            DO (ENTER-EG-IF-NEW-MAIL))

IA-M2 =    (AN INTERACTION-METHOD
            FOR CHECK-FOR-NEW-MAIL
            DO (KEY: 'EG/N' RETURN))
```

Fig. 7. Command Language Grammar (Moran 1981).

(a) The psychological view applies CLG as a model of an 'ideal' user's knowledge that shows the different kinds of knowledge that users have about systems. Moran, however, does not comment on the psychological validity of CLG as a model in this respect.

(b) The linguistic view uses CLG as a description of the structure of command language systems, which may be used to generate all possible 'command languages'. It should be noted that at the time of publication of CLG there was no uniform nomenclature of interaction styles (and, indeed, some currently well-known styles were not generally available), but CLG's claim is in principle valid for the description of all types of interaction mode.

(c) The design point of view applies CLG as a representation tool for specifying the system during the (top-down) design process to help the designer generate and evaluate alternative designs for the system.

However, only the third view of CLG as a description of the conceptual model is really worked out, and most prevalent. To this might be added that there are more powerful, or less cumbersome, grammars to describe the linguistic structure of an interaction language. Furthermore, Moran leaves us with only a number of suggestions about how a CLG representation might be analyzed to predict or evaluate aspects of the system's usability, such as performance times, memory load and learning. Sharratt (1987) presents some results of using CLG as a specification tool in a practical design exercise, in which he asked students to use CLG to specify a design for a transport time-table system. Sharratt concludes that CLG is useful for design specifications, but that it carries many of the drawbacks of a strictly top-down design process and leaves little room for design iterations. Furthermore, CLG cannot be used to describe the relation between the tasks and the information on the screen. CLG, however, seems to provide a valid framework to model (competent) users's knowledge – at least from the point of view of the system designer (Van der Veer 1990).

ETAG

ETAG or Extended Task-Action Grammar (Tauber 1988, 1990) is in many ways comparable with Moran's CLG. Both are techniques to describe the human–computer interface from the point of view of the user (the 'virtual machine'), both employ the notion of levels of interaction, and both use formalisms to specify the contents of, and the mapping between these levels. Tauber (1988) used Task-Action Grammar (Payne 1984) to describe how users have to rephrase their tasks in terms of lower level rules, until arriving at the physical actions submitted to the computer system. Tauber prefers to use the concept of 'basic tasks' (different from Payne's 'simple tasks'), defined as 'tasks for which the system provides a single command or equivalent unit of delegation' (Tauber 1988). The system's basic tasks should be distinguished from the user's 'unit tasks', defined as 'elementary primary tasks that may not be decomposed into other primary tasks' (Van der Veer 1990).

Whereas TAG only provides levels for purely notational reasons, ETAG uses CLG's well-chosen levels of abstraction, adding some refinements. This is done, because ETAG is also aimed at formally specifying the 'user's virtual machine', including the task-related semantics of the computer system. The user's virtual machine (UVM) is defined by means of a canonical basis, an ontology borrowed from Psycho-linguistics (Jackendoff 1983). Basically, the ontological or

Part of the canonical basis for a UVM

[CONCEPT}:: = [OBJECT] |[VALUE] |[PLACE] |[STATE] |[EVENT]

[PLACE]   :: = [place.IN ([OBJECT])] |[place.ON ([OBJECT])]
               |[place.ON-POS ([OBJECT]) |[place.ON-TOP ([OBJECT])]
               |[place.ON-TAIL ([OBJECT])]

[STATE]   :: = [state.IS-AT ([OBJECT], [PLACE])]|
               |[state.HAS-VAL ([OBJECT], ⟨ATTRIBUTE⟩, [VALUE])]

[EVENT]   :: = [event.KILL-ON ([OBJECT], [PLACE])]
               |[event.MOVE-TO ([OBJECT], [PLACE])]| ...

type[EVENT > event.MOVE-TO ([OBJECT: * o], [PLACE: * p])]
    precondition: [state.IS-AT ([OBJECT: * o], [PLACE: * p0])];
    clears: [state.IS-AT ([OBJECT: * o], [PLACE: * p0])];
    postcondition: [state.IS-AT ([OBJECT: * o], [PLACE: * p])];
end [EVENT]

A conceptual object and a conceptual event of a UVM

type [OBJECT > MESSAGE]
    supertype: [TEXT];
    themes: [HEADER], [BODY];
    relations: [place.ON-POS(1) ([MESSAGE])] for [HEADER],
                [place.ON-POS(2) ([MESSAGE])] for [BODY],
                [place.POSS-AT ([MESSAGE])] for [HEADER], [BODY];
    attributes: ⟨SENDER⟩, ⟨SENDING_ DATE⟩, ⟨RECEIVING_ DATA⟩,
                ⟨STATUS⟩, ⟨DELETION_ MARK⟩;
END [message]

type [EVENT > COPY_ MESSAGES]
    description: for {[MESSAGE: * x]}
                   [event.COPY-TO ([MESSAGE: * x],
                   [place.ON-TAIL ([MESSAGE_ FILE: * y]): * p2])];
    precondition: [state.IS-AT ([MESSAGE: * x],
                   [place.ON-POS.(i) ([MESSAGE_ FILE: * z]): * p1])];
    comments: 'copy messages x from file z onto the end of file y';
END [COPY_ MESSAGES].

A basic task from the dictionary

ENTRY 6:
[TASK > COPY_ MESSAGES],
[EVENT > COPY_ MESSAGES],
[MESSAGE_ FILE: * z],
T6[EVENT > COPY_ MESSAGES]
    [OBJECT > MESSAGE: (* x)][OBJECT > MESSAGE_ FILE: * y],
'copy messages from the current message file into a message file'.

Fig. 8. Extended Task-Action Grammar (Tauber 1990).

canonical basis describes the world in terms of concepts (such as: objects, places, and states), attributes, relations between objects, functions of objects (such as: object being at places, e.g., on top of others), and events which change existence, functions, and relations (such as killing, moving and copying objects).

The canonical basis indicates relevance for the user, and should be part of the user's virtual machine. In terms of Norman (1983) the UVM is the conceptual model of the target machine and, as such, equivalent to a competent user's mental model. In terms of Kieras and Polson (1985) the UVM describes the 'how it works' knowledge the user needs. An example of a conceptual event in an electronic mail system is to 'mark for deletion', which sets the attribute 'deletion mark' for an object 'message' that resides at a place in 'message file'.

The next level in ETAG consists of the dictionary of the basic tasks. This level lists which basic tasks are possible, and how they relate to the concepts of the UVM. Fig. 8 gives an example of the higher levels of an interface specification in ETAG. Fragments of the UVM of an electronic mail system are illustrated including part of the 'canonical' basis (a concept hierarchy) and a description of an object (a message) and an event. An entry of the dictionary of basic tasks shows the formal description of the semantics of a basic task (copy a message).

The dictionary of basic tasks corresponds to the top level of the production rules, which use the feature grammar of Task-Action Grammar to describe how to perform the basic tasks in terms of still lower levels, until the commands for the computer system are fully specified. ETAG employs a refinement of the levels of CLG to structure the process of derivation by introducing levels to specify the syntax, the referencing style (e.g., pointing versus naming), the lexicon and the keystrokes, respectively.

ETAG, although originally designed as a modelling tool for user interface design, has already been applied for modelling user performance and user knowledge (Van der Veer et al. 1990).

An evaluation of formal modelling techniques

Formal modelling techniques as analyzed in the previous section, are representation methods to specify aspects of human–computer interaction. Models represent what a user has to know or to do in order to accomplish tasks by means of a computer system. Models can be used to analyze the similarities and differences in the way tasks are to be done with and without a computer system, to evaluate usability characteristics of human–computer interfaces, to predict certain aspects of user behaviour, and to formalize the hierarchical design of user interfaces in terms of the knowledge of the user at multiple levels of abstraction.

At this point, one could ask to what extent the present formal modelling techniques can be used successfully for these purposes. and if not completely successful, what the main problem areas are in using these techniques. To answer this question, it is necessary first, to determine the special requirements the modelling techniques should fulfil to specify user knowledge for the afore-mentioned purposes. We explicitly mention 'special' requirements, because we are concerned with formal models for specification purposes, namely, of user knowledge for the purpose of task-environment analysis, knowledge analysis, performance prediction, and representation for design.

In the past, others have suggested various requirements, some of which have been adapted, either completely or partially, and some have been rejected. For example, from Green et al. (1988) we accepted the requirement that formal models should be usable for designers. However, when they write that 'The model must contain a representation of the external semantics' (p. 38), they notice an important problem area of the present models, but not an overall requirement. Also, we do not think that it is necessary that 'The model must describe a reasonably complete psychology' (p. 38), as long as the resulting inferences are valid and useful for the desired purpose.

What we consider important is that formal models are tools to represent user knowledge for the purpose of computer system design. In the introduction, Norman's (1983) notion of models in human–computer interaction was discussed to stress the point that the design of a system should be based on a conceptual model, derived from an analysis of the intended users, and the users' tasks, in order to attain mutual consistency between the system image, user's mental models, training materials, and the conceptual model. From this basis, four requirements for representation techniques for design purposes can be put forward.

(a) A conceptual model should be based on both the point of view of the user, and provide a complete and accurate representation of the design. Hence, formal models should provide a complete description at the different levels of abstraction, of the intended system.

(b) Representation tools should have a wide applicability. In computer systems design, a modelling technique is of little use when one has to resort to another technique, for example, just when the style of interaction is to change from one to another. Therefore, formal modelling techniques should be applicable to a variety of different kinds of users, styles of interaction, and types of tasks.

(c) Formal models are used as conceptual representations of computer systems to analyze and predict usability aspects, about knowledge, and about performance. Inferences from analysis and prediction are useful to the extent that they are valid. The same applies to the modelling techniques on which the inferences are based.

(d) Just like computer systems are mere tools for their users, formal modelling techniques themselves are tools to perform the tasks of their own users: the designers. Therefore, within the broader context of design, formal modelling techniques themselves should fulfil the requirements of being functional, easy to use, and easy to learn and remember.

The requirements we have selected apply foremost to the modelling techniques for design specifications, but only because the design specifications are proposed with the most general intentions. CLG, with its three views, for example is intended for design specification, but also for user performance prediction and knowledge analysis. On the other hand, the requirements also apply to the techniques with more limited aspirations. For example, to analyze user knowledge, the representation should be a valid one, and include all the relevant aspects of the user's knowledge in a variety of circumstances and be usable for the analyst. One may imagine, however, that more specific purposes demand a different relative weighting of the requirements.

The completeness of formal models

The requirement of completeness means that a formal modelling technique should enable a complete specification of the user interface at all the levels of abstraction involved in using and in designing the interface. The modelling techniques reviewed in this paper fulfil this requirement to a smaller or larger extent, but none completely covers the whole interface. According to Green (1990) completeness is not a key requirement, and it may be better to have a number of limited theories, each of which covers its domain of application well, than to have a few large theories which cover more questions but each question only to a limited extent. As such, it may be possible to employ Cognitive Complexity Theory to predict performance times, Task Action Grammar to evaluate the consistency of an interface and Command Language Grammar to specify the interface for design. In the practice of design however, it is preferred to deal with only a limited number of methods since otherwise usability problems might arise. Regarding completeness, the main omissions in the present modelling techniques can be found at the highest and at the lowest levels of abstraction.

At the higher levels, an analysis of the user's concepts of the external tasks and those of the device are either omitted, mentioned without further specification, or the user's goal task hierarchy is much more rigidly specified than this will be the case in reality. For example, in Reisner's Action Language only attention is paid to the syntactic and lexical aspects of the users actions, and no attention is paid to the semantics of the interface. Payne and Green's TAG theory goes a step further, and list the semantic features of tasks, such as that the so-called 'clipboard' is involved in a cut and paste action. However, they do not in any sense describe the nature of clipboards as a temporary storage place for data. Moran's CLG would contain a description of the clipboard, but it would probably need a comment to fully explain its nature. At the moment only Tauber's ETAG is able to give formal account of these semantic features because it uses an ontology. Recent work by Payne (1987) also deals with the semantics of the interaction, especially those of the device being used.

At the lower levels of abstraction generally a description is missing of the visual presentation of the interface, especially in relation with the state of the system. Regarding the TAG theory, Green et al. (1988) write that: 'it [TAG] does not exhibit the relation between actions and system display; as far as TAG is concerned, the VDU screen could have been turned off' (p. 30). Although Moran (1981) explicitly mentions a spatial layout level, it has not been specified any further, and therefore none of the present modelling approaches is able to address the presentation component.

The width of applicability of formal modelling techniques

The requirement of the width of applicability means that a formal modelling technique should be applicable to a variety of user-populations, types of tasks and ways of interacting with a computer. Here also, coverage is limited.

In the first place the knowledge or the performance of the user that is typically described refers either to the ideal user or to the competent user. The ideal user is taken to be one who has perfect knowledge and is only engaged in error-free and most efficient performance. On the other hand, the competent user is only perfectly knowing, but may commit performance errors. Even if the focus is on evaluating the interface as a whole, it is difficult to apply findings to real users, and especially to novice users, who may be characterized by their imperfect and even erroneous knowledge of the system (e.g., Briggs 1990). This point is not really problematic, because it refers only to the inability to predict task performance by individuals, whereas we are more concerned about evaluation of design and prediction of task performance in general. As an exception, CCT has been applied to model performance of non-expert users.

Secondly, except for Moran's ETIT analysis all the models essentially employ a context-free grammar or an equivalent method of representation, which means that performance on a given task is viewed independent of any other tasks and that only isolated tasks can be represented. The assumption that task performance is independent of for instance previous tasks does not seem to be in accordance with the reality of computer use.

The last point is the consequence of both the use of a context-free grammar, and the lack of a specification of the visual presentation component. Presently, it is not possible to apply the modelling techniques successfully to model other tasks than those requiring little or no control or revision of planning by the user during execution. In reality however, users do control their tasks based on the knowledge of delegation of other tasks (both in parallel and in sequence), and on the perception and interpretation of information on the screen and other system responses. Rassmussen (1987) points out that because of this restriction, formal models may eventually only apply to the least important and least interesting bits in human–computer interaction. Additional or alternative modelling techniques are required to address the dynamics in user task performance, like the PUMS approach (programmable user models) which covers problem solving to some extent (Young et al. 1989), or the Action Facilitation approach which is focussed on facilitating and inhibiting factors for task performance (Arnold and Roe 1987; Roe 1988).

The validity of formal models and modelling techniques

The third requirement of formal modelling techniques concerns the validity of the analyses and predictions delivered. This requirement does not only apply to task environment, knowledge analysis, and performance prediction. The representations for design purposes also carry a notion of what constitutes a good design, by the implicit choice to include certain features in the model as relevant and leave others out. Here, few problems can be mentioned; within the limited field of application for which the formal modelling techniques have been proposed, it has been shown that they indeed do what they are supposed to. To name a few, Reisner (1981) has shown that action language can be used to predict differences in ease of use between user interfaces. Card et al. (1983) and Polson (1987) mention a number of experiments in which user performance was predicted reasonably well by their respective models. The experiments described by Lerch at al. (1989) show experimental application of formal modelling to knowledge of commercially available systems, where both execution times and certain types of errors were succesfully predicted. Finally, Payne and Green (1989) describe an experiment in which Task-Action Grammar was successfully used to address subtle differences in the usability of interfaces, that other modelling techniques could not address.

There are several limitations connected to the afore-mentioned and other validation studies. The validation studies have generally used very simple user interfaces, they have almost always been performed by or under the supervision of the original authors of the method, and the studies have hardly ever taken place outside the research laboratory using full blown interfaces. Apart from establishing the utility of formal models by others than experts of the particular modelling technique, there is a need to establish the validity of the modelling techniques when used by non-experts, and a need to seek the limits of their applicability in real design.

The usability of formal modelling techniques

A final requirement formal models have to satisfy is being of use in the practice of interface design. Here, the relevant points are to what extent formal models can be used in all stages of design, the adaptation to other techniques used by designers, and the usability aspects of applying formal models.

As was argued before the present approaches to formal modelling do not cover all the aspects and levels of abstraction of user interfaces. The current techniques cannot completely specify the semantics of the computer system, the presentation component and other reactions of the system. Presumably, the most important factor in this is the lack of a specification of the semantics of the system, including the 'how it works' knowledge. because when the conceptual model of a design is known then it may be easily used, e.g., to guide the choice for a particular screen layout or for the contents of an error message. At present we can only say that work needs to be done in this area.

A second point is that formal modelling techniques are usable in design to the extent that they can be integrated with other techniques used by designers. As such, the use of formal models should both adapt to the very first stage of interface design, namely task analysis, and adapt to the final stages of design, such as software engineering, prototyping, and testing. In other words, formal models are tools to communicate very precisely about designs, but in rather abstract terms, which in some way or another have to be related to the real world. Formal modelling techniques may be expanded and adapted to the complementary methods that are applied in design. Summersgill and Browne (1989) report an attempt to integrate what they call 'functionality centred' and 'user centred' design techniques. Walsh (1989) notes that, although there is a gap between the techniques and notations of task-analysis, formal modelling, and software engineering. this is only a matter of a different focus, and not a matter of the inability to understand each other's language. Regarding the relation between formal modelling and task-analysis, what needs to be done is to find a way to translate the informal or semiformal representations of the tasks of users into a formal representation of the conceptual objects and operations which should be used in delegating tasks to the computer system. The relation between modelling user knowledge and software engineering is probably even less problematic, because both already use formal models. These models do, however, differ with respect to exactly what is modelled: aspects of the user or aspects of the interaction. Barnard and Harrison (1989) propose an interaction framework to model user–system behaviour as a bridge between modelling user knowledge and modelling system behaviour. Another promising development is the object-oriented approach, which may make it possible to create bridges between both task analysis and user knowledge modelling, and software engineering. Eventually, it may be possible to use the very same objects to represent the tasks and task entities of the users' task world, the conceptual objects and operations of the user interface, and the data objects and procedures of the computer system.

A third point concerns the usability of formal models for designers. The present formal modelling techniques are not easy to use and demand substantial effort. For example, Wilson et al. (1988) report that applying formal models does often require a high level of expertise on behalf of the designer. According to Sharratt (1987), who studied Command Language Grammar, this is especially the case when formal design specifications are changed, or design alternatives are to be compared. To attack this problem, there is a need for something like a designer's workbench, built around a particular formal modelling technique, or for providing facilities to employ different formal techniques. The most important thing the workbench should provide is a design approach to guide and structure design decisions. Furthermore, several tools will be necessary to relieve the designer from much of the administrative work, such as generating and changing formal design models (e.g., specialized editors that enable automatic semantic consistency checks, and templates for formalizing design attempts). Together, the design approach and the tool set should facilitate the integration between task analysis, formal modelling and implementation. The workbench should also provide for facilities to decrease expertise required in dealing with the formalism as such, by providing adequate on line help and explanation facilities during the different stages of design (see Van der Veer et al. (1990) for an example of a prototype of this type of tools).

Concluding remarks

In this paper it was argued that people use computers as mere tools to perform the tasks they have to do, and therefore the computer system should provide for the functions to enable users to delegate their tasks, and minimal effort should be required to learn and use the system. Furthermore, computer users create mental models of the system they are working with, which help them to explain the behaviour of the computer system, and serve to aid in planning their actions, thus reducing the mental effort required. It was argued that the designer of a computer system should consider the development of users' mental models, by taking a conceptual model as the basis for the design. Conceptual models can provide an accurate, consistent, and complete specification of the design, at different levels of abstraction in the knowledge of computer users. There are various, partially complementary methods to enhance a system's usability, each with specific advantages and disadvantages.

This paper focussed on formal modelling techniques to represent the knowledge users need to operate computer systems, which have the advantages of formality, early applicability, relative time and cost inexpensiveness, and usability as conceptual models. Various types of formal models were reviewed, according to their primary purpose: models for task environment analysis, models to analyze user knowledge, models to predict user performance, and representation models for design purposes. Formal modelling techniques should meet four special requirements, in addition to the general requirements imposed on formal systems and specification tools. For each of these requirements (completeness, wide applicability, validity and usability) the problems encountered with the present formal modelling techniques were discussed, along with possible solutions.

Regarding the question of what has to be done from now on, we can make a threefold distinction between addressing the limitations of the current models, the ongoing development of user-oriented computer systems design. and the use of formal modelling techniques in the design practice.

(a) The limitations of the present modelling techniques have to be addressed. Regarding the requirement of completeness, the present modelling techniques have to be extended, or alternative techniques need to he developed to enable (1) the specification of the visual presentation component, and (2) the specification of the semantics of the tasks and devices, in addition to task-action mappings. Where it concerns the requirement of a wide applicability, stronger modelling techniques have to be developed to address (1) the context sensitive aspects of the user–computer interaction, such as multi-tasking, and (2) the dynamic aspects of task control and planning, and the presentation of information and other reactions from the system. The requirement of validity demands validation studies to be done by independent researchers, using real computer systems outside the laboratory. Regarding the usability of formal models, opportunities should be investigated to bring about an integration between the techniques for task analysis, formal modelling techniques and software engineering methods. Also, tools should be developed to reduce the additional amount of effort and expertise, imposed by the present formal modelling techniques.

(b) Several remarks have been made about how to improve aspects of design approaches to increase the usability of computer systems. The question is now, how to develop a better design approach, which includes all improvements needed. On the basis of the previous discussion, it may be clear that we are convinced that a user-oriented design approach should be based on a conceptual model: an accurate, consistent and complete specification of the intended system, at the different levels of abstraction, and, most importantly, considering the knowledge of the intended user. A formal modelling technique can be chosen, either from this overview or from another source, on the basis of the four requirements previously discussed, and possibly additional requirements, such as opportunities to extend the model, or to integrate it with current design tools. Even before developing tools to facilitate the use of the modelling technique, (1) it should be determined if, and to what extent the technique and the resulting models can be integrated with task analysis and software engineering approaches. If such an integration is not possible, or only to a limited extent, then any other effort is useless. Otherwise, (2) an integration can be established, which is presumably not an easy undertaking. After the backbone of an overall design approach is thus created, (3) it may be refined, completed, and complemented with the required tools. Whereas the former two steps may take place as an entirely theoretical project, for the success of this last step practical experience will be inevitable, because only then the weak points and the gaps of the approach will show up.

(c) The present approaches to formal modelling have almost exclusively been used within the research domain of Cognitive Ergonomics. This has led to a situation where design has commonly be exemplified by small-scale studies in which either students, or even the authors, of a certain design method took part as designers of a computer system with maybe ten different functions to perform some artificial task. Although there are exceptions, the point will be clear: in order to bridge the gap between theory and practice of interface design, the application of formal models in actual interface design is required in order to gain new theoretical insight. Only by means of full scale design examples will it be possible to show to the design community the advantages of using formal models. In this respect, one successful computer system is more valuable than ten research papers on interface design. Both the development of formal modelling techniques and the development of methods to design usable computer systems are best served by practical experience.

References

Arnold, A.G. and R.A. Roe, 1987. 'User errors in human–computer interaction'. In: M. Frese, E. Ulich and W. Dzida (eds.), Psychological issues of human computer interaction in the work place. Amsterdam: North-Holland.

Backus, J.W., F.L. Bauer, J. Green, C. Katz, J. McCarthy, P. Naur, A.J. Perlis, H. Rutishauer, K. Samelson, B. Vauquois, J.H. Wegstein, A. van Wijngaarden and M. Woodger, 1964. Revised report on the algorithmic language Algol 60. A/S regnecentralen, Copenhagen.

Barnard, P. and M. Harrison, 1989. 'Integrating cognitive and system models in human computer interaction'. In: A. Sutcliffe and L. Macaulay (eds.), Proc. People and computers V. Cambridge: Cambridge University Press. pp. 87–103.

Briggs. P., 1990. Do they know what they're doing? An evaluation of word-processor users' implicit and explicit task-relevant knowledge, and its role in self-directed learning. International Journal of Man–Machine Studies 32, 385–398.

Card, S.K., T.P. Moran and A. Newell, 1983. The psychology of human–computer interaction, Hillsdale, NJ: Erlbaum.

Diaper, D., 1989. 'Bridging the gulf between requirements and design'. In: Proc. Simulation in the development of user-interfaces. Brighton, 18–19 May. pp. 129–145.

Foltz, P.W., S.E. Davies and P.G. Polson, 1988. 'Transfer between menu systems'. In: E. Soloway, D. Frye and S.B. Sheppard (eds.), Proceedings CHI '88. Amsterdam: North-Holland. pp. 107–112.

Frohlich, D.M. and P. Luff, 1989. Some lessons from an exercise in specification. Human Computer Interaction 4, 101–123.

Gould, J.D. and C. Lewis, 1985. Designing for usability: key principles and what designers think. Communications of the ACM 28, 300–311.

Green, T.R.G., 1990. 'Limited theories as a framework for human–computer interaction'. In: D. Ackerman, and M.J. Tauber (eds.), Mental models and human–computer interaction, Vol. 1. Amsterdam: North-Holland. pp. 3–39.

Green, T.R.G., F. Schiele and S.J. Payne, 1988. 'Formalisable models of user knowledge in human–computer interaction'. In: G.C. van der Veer, T.R.G. Green, J.M. Hoc and D.M. Murray (eds.), Working with computers: Theory versus outcome. London: Academic Press. pp. 3–46.

Innocent, P.R., M.J. Tauber, G.C. van der Veer, S. Guest, E.G. Haselager, E.G. McDaid, L. Oestreicher and Y. Waern, 1988. 'Representation of the user interface. Comparison of descriptions of interfaces from a designers point of view'. In: R. Speth (ed.), Research into networks and distributed applications. Amsterdam: North-Holland. pp. 345–359.

Jackendoff, R., 1983. Semantics and cognition. Cambridge, MA: MIT Press.

Karat, J., 1988. 'Software evaluation methodologies'. In: M. Helander (eds.), Handbook of human–computer interaction. Amsterdam: North-Holland. pp. 891–903.

Kieras, D., 1988. 'Towards a practical GOMS model methodology for user interface design'. In: M. Helander (ed.), Handbook of human–computer interaction. Amsterdam: North-Holland.

Kieras, D. and P.G. Polson, 1985. An approach to the formal analysis of user complexity. International Journal of Man–Machine Studies 22, 365–394.

Knowles, C., 1988. 'Can cognitive complexity theory (CCT) produce an adequate measure of system usability?' In: D.M. Jones and R. Winder (eds.), Proc. People and computers IV. pp. 291–307. Cambridge: University Press. Cambridge.

Lerch, F.J., M.M. Mantei and J.R. Olson, 1989. 'Skilled financial planning: The cost of translating ideas into action'. In: Proc. CHI '89: Wings for the mind. New York: ACM. pp. 121–126.

Moran, T.P., 1981. The command language grammar: A representation for the user-interface of interactive systems. International Journal of Man–Machine Studies 15, 3–50.

Moran, T.P., 1983. 'Getting into the system: External–internal task mapping analysis'. In: A. Janda (ed.), Proc. CHI '83. New York: ACM. pp. 45–49.

Murray, D.M., 1988. A survey of user cognitive modelling. National Physical Laboratory, NPL Report DITC 92/87.

Nielsen, J., 1986. A virtual protocol for computer–human interaction. International Journal of Man–Machine Studies 24, 301–312.

Nielsen, J., 1990. A meta-model for interacting with computers. Interacting with Computers 2, 137–160.

Norman, D.A., 1983. 'Some observations on mental models'. In: D. Gentner and A.L. Stevens (eds.), Mental models. Hillsdale, NJ: Erlbaum, pp. 7–14.

Oberquelle, H., 1983. 'On models and modelling in human–computer co-operation'. In: G.C. van der Veer, M.J. Tauber, T.R.G. Green and P. Gorny (eds.), Readings on cognitive ergonomics – Mind and computers. Berlin: Springer-Verlag. pp. 26–43.

Payne, S.J., 1984. 'Task action grammar'. In: B. Shackel (ed.), Proc. Interact '84. Amsterdam: North-Holland. pp. 139–144.

Payne, S.J., 1987. 'Complex problem spaces: Modelling the knowledge needed to use interactive devices'. In: H.J. Bullinger, B. Shackel (eds.), Proc. Interact '87. Amsterdam: North-Holland. pp. 203–208.

Payne, S.J. and T.R.G. Green, 1983. 'The user's perception of the interaction language: A two-level model'. In: Proc. CHI '83: Human Factors in Computing Systems. New York: ACM. pp. 202–206.

Payne, S.J. and T.R.G. Green, 1986. Task-action grammars: A model of the mental representations of task languages. Human–Computer Interaction 2, 93–133.

Payne, S.J. and T.R.G. Green, 1989. The structure of command languages: An experiment on task-action grammar. International Journal of Man–Machine Studies 30, 213–234.

Polson, P.G., 1987. 'A quantitative theory of human–computer interaction'. In: J.M. Carroll (ed.), Interfacing thought: Cognitive aspects of human–computer interaction. Cambridge, MA: MIT Press. pp. 184–235.

Polson, P.G., 1988. 'The consequences of consistent and inconsistent user interfaces'. In: R. Guindon (ed.), Cognitive science and its applications for human–computer interaction. Hillsdale, NJ: Erlbaum. pp. 59–108.

Polson, P.G. and D.E. Kieras, 1985. 'A quantitative model of the learning and performance of text editing knowledge'. In: L. Borman and B. Curtis (eds.), Proc. CHI '85. New York: ACM. pp. 207–212.

Preece, J. and L. Keller (eds.), 1990. Human–computer interaction: Selected readings. London: Prentice-Hall.

Rassmussen, J., 1987. 'Cognitive engineering'. In: H.J. Bullinger and B. Shackel (eds.), Proc. Interact '87. Amsterdam: North-Holland. pp. xxv–xxx.

Reisner, P., 1981. Formal grammar and human factors design of an interactive graphics system. IEEE Transactions on Software Engineering 7, 229–240.

Reisner, P., 1983. 'Analytic tools for human factors of software'. In: A. Blaser, M. Zoeppritz (eds.), Proc. Enduser systems and their human factors. Lecture Notes in Computer Systems no. 150. Berlin: Springer-Verlag, pp. 94–121.

Reisner, P., 1984. 'Formal grammar as a tool for analyzing ease of use: Some fundamental concepts'. In: J.C. Thomas and M.L. Schneider (eds.), Human factors in computer systems. Norwood, NJ: Ablex. pp. 53–78.

Richards, J.N.J., H.E. Bez, D.T. Gittins and D.J. Cooke, 1986. On methods for interface specification and design. International Journal of Man–Machine Studies 24, 545–568.

Roberts, T.L. and T.P. Moran, 1984. The evaluation of text editors: Methodology and empirical results. Communications of the ACM 26, 265–283.

Roe, R., 1988. 'Acting systems design – An action theoretical approach to the design of man–computer systems'. In: V. de Keyser, T. Qvale, B. Wilpert and S.A. Ruiz Quintanilla (eds.), The meaning of work and technological options. New York: Wiley.

Rohr, G. and M.J. Tauber, 1984. 'Representational frameworks and models for human–computer interfaces'. In: G.C. van der Veer, M.J. Tauber, T.R.G. Green and P. Gorny (eds.), Readings on cognitive ergonomics – Mind and computers. Berlin: Springer-Verlag. pp. 8–25.

Sharratt, B.D., 1987. 'Top-down interactive systems design: Some lessons learnt from using command language grammar'. In: H.J. Bullinger and B. Schackel (eds.), Proc. Interact '87. Amsterdam: North-Holland. pp. 395–399.

Shneiderman, B., 1982. Multiparty grammars and related features for defining interactive systems. IEEE Transactions on Systems. Man and Cybernetics SMC-12, 93–133.

Simon, T., 1988. 'Analysing the scope of cognitive models in human–computer interaction: A trade-off approach'. In: D.M. Jones and R. Winder (eds.), Proc. People and computers IV. Cambridge: Cambridge University Press. pp. 79–93.

Smith, S.L. and J.N. Mosier, 1984. Design guidelines for user–system interface software. Bedford, MA: The Mitre Corporation.

Summersgill, R. and D.P. Browne, 1989. Human factors: Its place in system development methods. Sigsoft 14, 227–234.

Tauber, M.J., 1988. 'On mental models and the user interface'. In: G.C. van der Veer, T.R.G. Green, J.M. Hoc and D.M. Murray (eds.), Working with computers: Theory versus outcome. London: Academic Press. pp. 89–119.

Tauber, M.J., 1990. ETAG: Extended Task Action Grammar – A language for the description of the user's task language'. In: D. Diaper, D. Gilmore, G. Cockton and B. Shackel (eds.), Proc. Interact '90. Amsterdam: North-Holland. pp. 163–168.

Van der Veer, G.C., 1989. Visual aspects of user interfaces – A psychological view. Notizen zu Interaktiven Systemen 17, 19–27.

Van der Veer, G.C., 1990. Human–computer interaction: Learning, individual differences, and design recommendations. Ph.D. Thesis, Free University, Amsterdam.

Van der Veer, G.C., F. Yap, D. Broos, K. Donau and M.J. Fokke, 1990. 'ETAG – Some applications of a formal representation of the user interface'. In: D. Diaper et al. (eds.), INTERACT '90 – Third IFIP Conference on Human–Computer Interaction. Amsterdam: North-Holland.

Vossen, P.H., S. Sitter and J.E. Ziegler, 1987. 'An empirical validation of cognitive complexity theory'. In: H.J. Bullinger and B. Shackel (eds.), Proc. Interact '87. Amsterdam: North-Holland. pp. 203–208.

Walsh, P., 1989. 'Analysis for Task Object Modelling (ATOM): Towards a method of integrating task analysis with Jackson System development for user interface software design'. In: D. Diaper (ed.), Task analysis for human–computer interaction. Chichester: Ellis Horwood.

Whitefield, A., 1987. 'Models in human computer interaction: A classification with special reference to their uses in design'. In: H.J. Bullinger and B. Shackel (eds.), Proc. Interact '87. Amsterdam: North-Holland. pp. 57–63.

Wilson, M.D., P.J. Barnard, T.R.G. Green and A. MacLean, 1988. 'Knowledge based task analysis for human–computer systems'. In: G.C. van der Veer, T.R.G. Green, J.M. Hoc and D.M. Murray (eds.), Working with computers: Theory versus outcome. London: Academic Press. pp. 47–87.

Van Wijngaarden, A., B.J. Mailloux, J.E.L. Peck and C.H.A. Koster, 1969. Report on the algorithmic language ALGOL 68. Numer. Math. 14, 79–218.

Young, R.M., T.R.G. Green and T. Simon, 1989. 'Programmable user models for predictive evaluation of interface designs'. In: K. Bice and C. Lewis (eds.), Proc. CHI '89. New York: ACM. pp. 15–19.

Acta Psychologica 78 (1991) 69–96
North-Holland

Errors and theory
in human–computer interaction

Paul A. Booth *

University of Salford, Salford, UK

The problem of human–computer interaction (HCI) is an issue of, not only matching system functionality to the needs of the user in a specific work context, but also a question of presenting an image of the system that can be easily understood. Attempts to support design in such a way so as to overcome these problems are characterized as having moved through the 'guidelines' approach of the 1970s, to an 'analytic methods' perspective in the 1980s. However, existing approaches account for only limited aspects of interaction, and areas such as the misunderstandings (errors) that users experience have received less attention. In addition, attempts to construct methods that might allow designers to analyse and understand user errors have met with only limited success. It has been suggested that the theories and constructs embodied within some of the analytic methods might be of use (although such claims have not been made by the originators of these methods). We consider the failure of a usable tool (ECM), and the limited HCI theories embodied in TAG and GOMS, to adequately account for three example errors. This failure, it is argued, suggests that the next stage is for those within the HCI field to develop broad theories of understanding and action, that can form the basis for design tools, and be communicated to designers.

Introduction

In the past decade the importance of human–computer interaction has been generally recognized. Recently, the USA-based Association of Computing Machinery listed human–computer communication as one of its nine core areas for a computing degree (Denning et al. 1989). Human–computer interaction may have been presented under a variety of names (man–machine interaction, man–machine systems, human–computer interface, etc.), but the core problems remain the same, although approaches change.

* I am grateful to Gill Brown, John Dowell and Erik Hollnagel for their comments on an earlier draft of this paper and for some of the better ideas expressed within it.

Author's address: P.A. Booth, Dept. of Mathematics and Computer Science, University of Salford, Salford M5 4WT, UK; E-mail: PABooth@mcs.salford.ac.uk.

The HCI problem

For convenience we can draw a distinction between the problems of functionality and understanding in HCI. The first problem is a question of identifying the facilities the user requires within a particular organizational framework, while the second is concerned with presenting facilities in a way that is comprehensible to the user. These problems appear to have a common cause: design team members use themselves as models of the user (Alty and Coombs 1980; Eason 1976), yet the gulf between the designer and the user is considerable – software engineers, technical authors, etc., are experts in a technical field and novices in the user's area of expertise. Users and designers apply different criteria when assessing a system; they have different goals and hence perceive a system differently (see Hiltz 1980). Part of this perception is based around the functionality of the system, but a large part is related to how users and designers understand a system differently.

Disentangling the relationship between functionality and understanding is not easy. The way a user understands a system mediates the user's perception of functionality. Typically, users learn only a few commands and fail to explore all of the system (Eason 1984). This is not only true of the banking systems that Eason considered, but also of safety-critical systems such as process plants and the like. Moreover, even when users are fully aware of the functionality of a system, misunderstandings can be critical. For example, at least part of the problem which led to the 'Kegworth disaster', where in January 1989 a Boeing 737-400 crashed into a major UK highway, was that the pilots misunderstood (or did not trust) vibration indicators and shut down the one engine that was still functioning correctly (Gavaghan 1990). Of course, as with any major accident, there are many contributing causes. This misunderstanding was not the only cause, and a further problem was that the co-pilot was distracted for over two minutes while he reprogrammed the flight management system to allow them to divert to the airport they eventually failed to reach – a system that seems to have been designed for normal operation, but not for emergencies.

Clearly, misunderstandings or misinterpretations of systems can have serious consequences in safety-critical environments, but how users understand systems is also important for productivity and for continued use of a system in all environments. Users will persevere with a system for only a limited period of time (a 'honeymoon' period), and many systems are left to collect dust after being rejected, while others are only partially used. These examples, however, may appear as anecdotal evidence – how widespread is this problem?

How do we know that there is a problem?

Many computer systems fail, not because they do not work technically, but because their users choose not to use them (Eason 1988). Companies involved with developing and supplying information technology (IT) have little to gain by publicizing such failures, and the extent of the problem is often not fully understood outside the IT industry. For example, a study of IT implementation in small businesses revealed a success rate of only 40% (Wroe 1986), while other studies have suggested an even lower success rate (Hoerr et al. 1989). The picture is often worse if large companies are considered (Mowshowitz 1976), while specialist areas such as manufacturing technology (Majchrzak and Roitman 1989) or decision-support systems (McCosh 1984) fare no better. Even well-accepted technology, such as word-processing systems, show an unacceptably high failure rate (Pomfrett et al. 1984). The results of these failures can be costed in millions of dollars: one example is the US service industry – despite IT investment in excess of $100 billion in the last ten years there has been little if any productivity growth (Bowen 1989; Hackett 1990). Much of this can be attributed to systems that either were not used at all or were not used as they had been intended. Many of these failures are directly related to poor interactional characteristics. In short, systems are frequently confusing and difficult to learn, even so-called 'friendly' systems can be shown to create confusion (Jones 1989). Consequently, the way users understand systems is a key issue, and the issue that cognitive theory needs to address. The developing discipline of HCI has responded to this challenge, but in a variety of ways.

Craft approach. During the late 1960s and 1970s HCI developed as a craft discipline, to use Long and Dowell's (1989) term. Practitioners, using their backgrounds in ergonomics, computing and psychology, developed a deep knowledge of HCI problems through acting as consultants to various projects, or as trouble-shooters within large IT companies. The disadvantage of this approach was its unstructured and informal nature. When practitioners disagreed there was no means of choosing between their conflicting advice – this advice was based upon experience and not upon any mutually acceptable set of axioms. Moreover, the informal nature of this craft discipline meant that training new practitioners was a difficult process. In short, the knowledge possessed by these practitioners was destined, at least in part, to die with them.

Guidelines. The answer to this problem, adopted in the 1970s and early 1980s, was to embody this knowledge in the form of recommendations (Carroll and Thomas 1982), design rules (Gaines 1981), or guidelines, as they are more commonly called. Gilmore et al. (1989) have suggested several hundred guidelines, while Smith and Mosier (1984) have produced guidelines in their thousands. There are, however, problems with guidelines (Booth 1989; Thimbleby 1990). Many guidelines are vague to the point of being meaningless. Even when guidelines are detailed and well-informed (e.g. Marshall et al. 1987) they tend, nevertheless, to be ignored by designers in the inevitable rush to meet deadlines.

Analytic methods. A more rigorous approach has been to employ analytic methods (also called formal methods, user modelling techniques or cognitive grammars) to model either the system or the user (Barnard and Harrison 1989). Techniques for modelling the system have been aimed at identifying formal inconsistencies, etc., while cognitive modelling techniques have attempted to apply limited, approximate cognitive theories (Green 1987; Card et al. 1983). This approach has been popular for most of the 1980s, since Card et al. (1983) set out GOMS and Reisner (1981) suggested using BNF. While there have been many subsequent developments of this approach (e.g. Kieras and Polson 1985; Payne and Green 1986; see Grant and Mayes 1991 for a review), a number of drawbacks have been identified (Booth 1991; Briggs 1989; Carroll 1985; Détienne 1990; Green 1987; Knowles 1988). Their areas of applicability; the goals set for these methods have been progressively re-defined (see Moran 1986) to encompass less of the design process. Although a number of theoretical objections to these methods have been raised, it has been the practical problems of their apparent complexity (Bellotti 1988). Even when designers can understand these methods there is, nevertheless, a perception of a need for a large effort for a small reward – that has led to the belief in some quarters that design teams are not going to employ these methods as they currently exist.

The future. It appears, then, as though system and user modelling techniques are likely to acquire a status similar to that now afforded to the guidelines approach – they are useful but limited. Nevertheless, guidelines and the modelling methods may eventually play some small part in a wider solution, but they are not the answer to the HCI problem. What, then, of the future?

The forces that are likely to shape development in HCI in the 1990s may be many. This having been said, we can certainly identify some of the emerging trends. One is that the overall orientation adopted within HCI will be less towards prescriptive design, and more towards *enabling* creative, but user-centred design (see Guindon 1990). As part of the movement towards examining the realities of design, rather than concentrating upon design ideals, there is also likely to be more emphasis on design failures (Brooks 1990). A design failure can be thought of in a number of ways. In one sense a failure can be defined organizationally (Eason 1984). It can be considered as a mismatch between the working practices of groups within an organization and the nature of the facilities provided by the system. That is to say that the functionality and intelligibility of the system can be acceptable, but the software has not been designed to take organizational factors into account, such as conflicts between groups, existing practices, etc. (see Bjørn-Andersen et al. 1987). At another level it can be considered in terms of tasks, or functionality. The facilities the system provides, although intelligible, are not what users require. Finally, we can consider design failures as aspects of systems that cannot be easily understood by user – sometimes referred to as errors (or misunderstandings).

Design failures, when defined in the larger sense (i.e. whole systems failing rather than aspects of the system failing), have tended to receive less attention than other areas. Similarly, design failures when considered at a more detailed level (i.e. errors, misunderstandings, dialogue failures) have been paid little attention in the HCI field when compared with analytic methods and the like. Indeed, within industry the area can be something of a taboo subject. Although errors are often considered during design, they are treated as secret product information – something to be hidden from users and competitors.

Errors as an example

Previous papers have considered the nature of a simple technique for analysing errors (Evaluative Classification of Mismatch, ECM: Booth 1990d), the experience of a design team using the technique (Booth 1990c), and limited evidence that the application of the technique improves design (Booth 1990b). However, the theoretical limitations of this technique, and its consequent piecemeal analysis, have been its greatest drawback. The question we will address here is that of whether the constructs and theory embodied within two analytic methods can make up for these limitations.

Despite the emphasis within HCI on guidelines and formal techniques there have been some attempts to consider errors (e.g. Barnard et al. 1982, 1989; Carroll and Carrithers 1984; Davis 1983a b; Janosky et al. 1986; Lewis and Norman 1986; McMillan and Moran 1985; Monk 1986; Norman 1983; Riley and O'Malley 1984; Roberts and Moran 1983; Thimbleby 1982; Welty 1985; Young and Hull 1982). Recently, however, there has been an increasing recognition that errors are an important topic for study (e.g. Bannon and Bødker 1989; Bradford et al. 1990; Brodbeck et al. 1989; Buckley and Long 1990; Cuomo and Sharit 1989; Doyle 1990; Frese et al. 1990; Harding and Rengger 1989; Norman 1988; Reisner 1990; Rouhet and Masson 1990). Indeed, there appear to have been as many papers on errors in HCI published in the last year as have been published in the previous ten years.

Why look at errors?

Errors have often been the subject of humorous anecdotes, or less often the basis for evaluation during experimental assessments. More importantly, however, errors are frequently considered during design, where the misunderstandings that occur between users and systems are used to inform redesign during prototyping. Such prototyping is less often an opportunity for users to 'design, redesign and refine...' systems as Clark et al. (1984) had hoped, but provides an opportunity for a design team to observe users. The importance of considering the errors that users make in laboratories is that, providing the scenarios they are provided with take them through a realistic portion of the functionality of a system, these errors will cause users problems once the system is in the field if they remain unaddressed (Marshall et al. 1990). It is these problems that can often lead users to abandon attempts to understand and use the system – not just a case of user satisfaction, but an issue that has commercial and economic implications.

Effecting change

Of course, it is possible to argue that significantly changing a system during design in any major respect is not possible because the outcomes of any project are determined before the project begins (Curtis 1990). In many respects this perspective may be accurate. The way a system fits into an organization, and the functionality the system is likely to provide, have frequently been specified (even if not formally or even consciously) beforehand, and are hard to change in any substantial way. Design team members have much invested in their designs and changing a large part of a system involves a change in *beliefs* on the part of design team members – *beliefs* about what functions are required and how they should operate, *beliefs* about how users will interpret what they see on the screen. Some may object to the term *belief*, yet many design requirements are nothing more than articles of faith. Consequently, changes in belief require designers, not to be told that this or that is wrong, but to *witness* the falsity of their beliefs. It is for this reason that users are frequently video-taped during testing sessions – so that all members of the design team can see the crucial segments where misunderstandings become clear, and where new or altered functions are requested.

Consequently, *it is possible to change designs*, although we might still accept the general tenet of Curtis's (1990) observations – many aspects of a design are almost impossible to change without unrealistic investment. It is important to note that those aspects of a design that can be changed are those that concern how the system is presented to users – the features that mediate the user's interpretation and understanding of the system. Consequently, this involves the analysis of error. But what do we mean by the term *error*?

What is an 'error', and what do we know about them?

Rasmussen (1985) points out that identifying an error is sometimes a difficult question of judgement, while Lewis and Norman (1986) question whether error is the correct term at all. The term appears to assign blame. More important objections are raised by Hollnagel (1983) where he points out that 'errors' are actions – they are not special actions with different mechanisms, but the product of the same cognitive system that produces so many 'correct' or 'appropriate' actions. The significance of this observation is that we cannot understand error without understanding action.

We do know, however, that inappropriate actions – those we label as 'errors' after the event, cause problems. Carroll and Carrithers (1984) found in their study that users spent 25% of their time recovering from errors, while experience with other systems suggests that this percentage may actually be lower than the average, as users of other systems spend even more time recovering from errors. Moreover, once misunderstandings occur, they tend to 'snowball' (Janosky et al. 1986), sometimes leading to a strong emotional response on the part of the user (Brodbeck et al. 1989; see Booth 1990a, for a slightly more detailed account of the drawbacks of errors).

Of course, errors can be useful when they help to direct the user to a more appropriate understanding of a system. Papert's (1980) Logo was built upon the idea of constructive breakdowns, as have more recent attempts to construct learning environments (e.g. Howes and Payne 1990). Indeed, errors can be seen as a prerequisite to learning (Bannon and Bødker 1989).

We can, consequently, distinguish two views of errors: they are damaging occurrences that cause immense disruption and even distress to users; and they are invaluable points of breakdown that direct users towards more appropriate representations of the systems they are using. But which is correct? The answer, of course, is that they are both correct. Errors can be helpful or disruptive – it all depends upon *(i) the ease with which the user can recover from an error; and (ii) the extent to which the system provides cues or features that productively direct the user towards a more appropriate understanding.*

Types of errors. But what sorts of errors are we considering here? Situations where we are learning about a system involves the development of a representation of some sort. Misunderstandings, mismatches, misrepresentations, mistakes, or whatever else we might choose to call them are conceptual errors. They occur more often during initial periods of learning. There is, of course, another type of error – errors where the user (or operator) understands a situation well but slips or lapses into an inappropriate routine. These are sometimes called human errors, although the term is also applied to cover all errors, including conceptual errors. It might, however, be more accurate to call these automaticity errors – where we slip into a well-known automatic routine. It appears as though, at least within HCI, conceptual errors cause many more problems than automaticity errors, especially if recovery from automaticity errors (i.e. sophisticated 'undo' facilities) is easy (although this may be less true in safety critical systems). A further point to consider is that automaticity errors are only likely to affect those who have had some considerable experience with a system (or other computer systems).

Classifying errors

Many schemes have been suggested for analysing errors. Almost without exception, all involve the same approach of classifying the error. Errors can be classified according to whether they occur at a skill, rule or knowledge-based level (Rasmussen 1983, 1986); whether they are slips or lapses (automaticity errors) or mistakes (conceptual errors) (Norman 1983); and according to whether the error occurs at a task, semantic, syntactic or interactional level (Moran 1981; see Davis 1983a, b). Others have suggested a more explicitly linguistic approach, and Goodman (1987) proposes a comparatively large taxonomy of dialogue confusions, while Harding and Rengger (1989) have proposed a problem description language.

The need for the Evaluative Classification of Mismatch (ECM) scheme has been argued elsewhere (Booth 1990b, c, d). However, the key points are that the ECM scheme is usable by a design team; it is user-oriented; and it identifies the exact points in the design that are inappropriate from the user's perspective. This last point, and to a lesser extent the point regarding the usability of the scheme, distinguish it from other less design-oriented classification schemes. One problem, however, is that the scheme, although practically useful, is theoryless. It offers no greater understanding of the user's representation – it is only a terminology; a system-independent structure that helps designers to think through the user's problems. The question we will address here, *en route* to considering the nature of HCI itself, is *can existing HCI*

	Object	Operation
Concept	object-concept mismatch	operation-concept mismatch
Symbol	object-symbol mismatch	operation-symbol mismatch

Fig. 1. The four classes into which a mismatched element should fit.

theory make up for this shortfall? Can the limited theories embedded within formalisms such as Task Action Grammar (TAG – Payne 1984; Payne and Green 1986) and Goals, Operators, Methods and Selection Rules (GOMS – Card et al. 1983) provide a useful means by which we can understand users' misunderstandings (although no such claims were made by the originators of these methods)? First, however, we need to briefly describe ECM, as well as a little of the background of both GOMS and TAG.

The ECM scheme

The approach adopted within ECM is one where classification is based upon the user's perspective. What is *right* and *wrong* in the system relies entirely on the user's view of the system and task – if the user does not understand an aspect of a system, then it is the system that is assumed to be 'wrong', not the user. The four possible classes of mismatch following the identification of a dialogue failure can be seen in fig. 1. The classification process entails four stages through which the proposed classification process should proceed (see fig. 2).

Stage 1: Identifying a dialogue failure. The first stage of the classification process involves the identification of a representational mismatch or user-system error. In practical terms, this means the detection of a failure in the dialogue between the system and the user (for an explanation of the criteria see Booth 1990d).

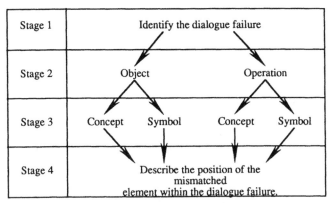

Fig. 2. The process of classification.

Stage 2: Identifying an object or operation. Having identified a dialogue failure then the next stage is to identify the object or operation associated with the failure. *Objects:* A data file in a system, an applications package, a figure, character or window on a screen might all be considered to be objects. *Operations:* An operation *is an action which is performed upon an object or objects within the system.* Saving a file, deleting a character, changing a shape in a graphics package, creating text in a word-processing package are all operations. A general rule of thumb for distinguishing objects from operations is that an operation is something that is done to an object whereas an object has operations performed upon it.

Stage 3: Identifying the mismatch type. Once the dialogue failure and the associate (an object or operation) of the dialogue failure has been identified then the next stage is to classify it regarding its cause. To do this, either the concept or the symbol of the object or operation is identified as being at fault. *Concepts: A concept may be either an object or operation whether represented mentally (the user) or computationally.* A concept mismatch is a *fundamental difference in the understanding and representation of system or task objects or operations.* *Symbols:* The term *symbol* is taken to mean *a word, character, sign, figure, shape or icon employed by either the user or the system to represent an object or operation within the system.* A symbol mismatch is not one of fundamental understanding, *but occurs where the computer and the user adopt different terms or visual images to represent the same concept.* A rule of thumb for distinguishing symbol from concept mismatches might be whether the user's symbols for the systems objects and operations easily map onto the designer's version of the task or system.

Stage 4: Positioning a mismatched element. Once a dialogue failure has been identified, and its causes classified, these causes (or mismatched elements) need to be described with respect to their position within (or contribution towards) the dialogue failure. The crucial question that is addressed at this stage in the classification process is: *what role did the element play in the dialogue failure?* The important point about distinguishing mismatches is that to properly classify a dialogue failure the observer must not only discover what happened (i.e. which operations were applied to which objects) but must also elicit the user's view of the task and system. The physical actions and consequences may remain constant, but the classification depends upon the user's view of the task and system.

GOMS and TAG

Card et al.'s (1983) approximate theory requires a task to be decomposed into its goals and subgoals, the operations required to achieve these goals, different methods for achieving goals, and rules for choosing between alternative methods. This scheme can be used to predict routes through tasks, as well as the time to complete a task. Although there is some evidence that users do not make the rational choices between alternative methods that GOMS assumes (MacLean et al. 1985), the attraction of this sort of production-rule theory is such that it has been used as the basis for more advanced methods (i.e. Kieras and Polson 1985). Nevertheless, the drawbacks of GOMS are that it assumes expert and error-free performance, while providing limited information (i.e. time to complete a task and the route through the task).

Payne and Green's (1986) TAG is based on notation very similar to Backus Naur Form (see Reisner (1981) for a discussion of the potential application of BNF to HCI). The central notion of TAG is that we, as users, have a library of cognitively automated *simple tasks* that we perform routinely. We can combine these simple tasks in various ways to achieve different effects. The ordering of these simple tasks is determined, in Payne and Green's theory, according to rule schemata (see Booth (1989), Payne and Green (1986) for an explanation). To analyse a system using TAG we first identify the system's concepts and then simple tasks. The number of different orderings of these simple tasks, for the different concepts, is represented in the number of rules required to describe the system. Essentially, the more rules that are required then the more complex, and possibly less consistent, the system. TAG, then, provides a measure of complexity. However, although TAG has certain theoretically attractive features (i.e. such as the way in which errors can be explained) it is considered to be difficult to learn, and gives small system inconsistencies as much prominence as large inconsistencies (Green 1987).

Although a number of researchers have identified both theoretical and practical drawbacks with this approach (Bellotti 1989; Booth 1991; Briggs 1988; Carroll and Rosson 1985; Green 1987; Knowles 1988) there is no a priori reason why the theoretical constructs and terminology provided by these methods might not be used to fruitfully classify and understand errors. Indeed, using these theories in this way overcomes some of the theoretical and practical objections that have been previously raised.

Representational mismatches – Some examples

Here, some of the descriptions of misunderstandings provided by Jones (1989) have been used. We will first analyse them using ECM, and then using GOMS and TAG. Although ECM has a number of drawbacks, as mentioned earlier, it nevertheless provides a baseline against which we can judge the usefulness of GOMS and TAG. Firstly, however, we need to outline some of the background to the observations made by Jones (1989).

Jones asked a number of users, ranging from people who had no computing experience to some who had several years experience of using IBM PCs, to use the Apple Macintosh PC and its manual. Jones observed, and to some extent interpreted, users' actions and the misunderstandings that occurred. Indeed, it is the interpretative aspect of Jones's descriptions that makes them so interesting and useful. We will examine just three of his problem descriptions.

Problem descriptions

Problem 1. In the Macintosh desktop system certain icons (i.e. folders or disks) can be opened to form windows on the screen (see fig.

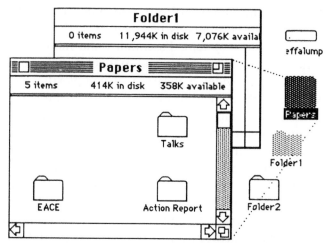

Fig. 3. A diagram showing a hollow disk icon and its window (the dotted lines have been drawn in to indicate the relationship), a hollow folder and its window and a second folder that is closed.

3). The user, however, did not understand the relationship between icons and windows. Nor did this user understand what was meant in the manual when it stated that an icon becomes *hollow*. For those who do not have any experience of the Apple Macintosh, it may be useful to describe this phenomenon. An icon may represent a folder or disk which contains files, applications, etc. When this icon is opened (usually by placing the mouse pointed over the icon and pressing the mouse button twice) a window appears (see fig. 3). The icon, however, remains where it is, but turns from being generally white to a darker shade of grey. It is in this state that the icon is considered to be *hollow*. This hollow icon now has new properties. It cannot be opened of course, because it is already open. If, however, the open operation is performed on it (i.e. the mouse pointed is placed over the icon and the mouse button pressed twice) this will have the effect of bringing the icon's window to the front of any other windows that are open on the desktop.

Problem 2. When a disk is ejected from the computer its icon, and the icons of any of its files on the desktop, do not disappear, but become hollow (see fig. 4). In this situation, however, if there is an attempt to open any of the icons this will not immediately cause them to either open nor their windows to come to the front, but will cause the system to ask for the disk to be reinserted. A user complained that

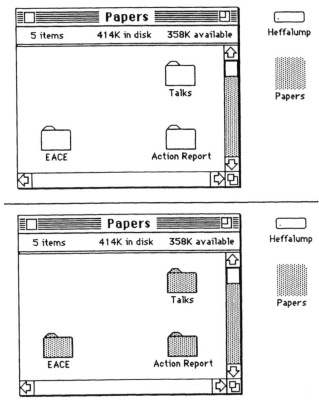

Fig. 4. A diagram showing the hollow disk item and its window before ejection and after ejection. Notice that the folders in the disk window become hollow after the disk has been ejected.

he thought the icons should disappear when the disk was ejected. To get rid of the icons they have to be placed in the wastepaper icon (see fig. 4). It is actually possible to eject a disk and remove its icons in one action, and this involves placing the disk icon directly into the waste-paper icon. Many users see this as counter-intuitive, however, and fear that their disk might be erased.

Problem 3. This final problem is related to the operations the Macintosh will perform when changing between applications. As the Macintosh loads an application (actually the wordprocessing application MacWrite in this case) it requires the user to swap a number of disks. The Macintosh will read data from one disk, eject it, and ask for the other disk to be inserted. After reading data from this one it then

ejects this disk and asks for the previous disk to be inserted. This cycle
can continue for as many as seventeen or more swaps. Most of Jones's
subjects believed that there was something wrong, and that they had
acted incorrectly in some way. Subjects did not realize that if they
continued to swap the disks the application would eventually appear
for them to use. Many of these subjects abandoned the session and
gave up when confronted with what they perceived as an insurmounta-
ble problem.

Analyses: ECM

Problem 1. This first problem, to recapitulate, was where the user
could not understand how an icon could open to become a window.
This object, this folder or disk icon that can open into a window, has
properties that the user did not expect. It is the object that is mismatched, and it appears to be the concept of the object that is at fault
from the user's perspective. Either the system needs to be changed in a
way that is more commensurate with users' expectations or the user
needs to be informed in some way of the nature and properties of
icons. This, however, is a design question. As we are concerned with
only analysing and understanding these problems rather than solving
them, we will not pursue these redesign issues.

Problem 2. This second problem was where the icons associated with
a disk did not disappear when the disk was ejected, contrary to the
user's expectations and wishes. We could argue that this is clearly a
problem with the operation of the system, as the system does not
operate as the user expects, and that it is the concept of the operation
that is mismatched with the user's representation. This, then, is an
operation-concept mismatch. On the other hand, as it is possible to
eject a disk and remove its icons from the desktop by placing the icon
in the wastepaper basket (i.e. as the operation the user wants is
possible) then we might characterize this as an operation-symbol mismatch, as the operation the user requires is present, but not in a form
the user recognizes.

The practical problems of agreeing classifications using ECM, evident here and mentioned earlier, are typical of the dilemmas that faced
the design team who used ECM to analyse problems and guide their
re-design discussions (Booth 1990c). It is possible to argue for more
than one classification of a problem depending upon the perspective

that is adopted when considering the problem. Although such problems tend to occur in a minority of cases, they are, nevertheless, significant.

Problem 3. In the final problem, where the Macintosh required users to swap disks a large number of times and users thought that had made some mistake, we can classify this as an operation-concept mismatch. This is because the notion of the operation that is being performed, that of loading the data needed for the application in blocks from different disks, is not shared by the user and the system.

Analyses: GOMS

In attempting to classify the problems in accordance with structure of the GOMS scheme we will attempt to slot representational mismatches into the four central constructs of Card et al.'s (1983) approximate theory. To do this we will identify whether it was a goal that was mismatched between the representations held by the user and the system; whether it was an operator that was mismatched; whether it was a method (series of operators) that was mismatched; or a selection rule that was not mutually shared.

Problem 1. In the case of the first problem, where icons open to become windows and the user did not understand this, we might argue that it is a goal mismatch. This is because the user did not want objects on the screen to open into windows. The problem is, of course, *the goals at which level?* Overall goals of producing a piece of text, etc., are most probably well-matched. Consequently, we might argue that it is not a goal mismatch, but the way in which the goal is achieved. Is it the operator or method that is mismatched? It appears to be difficult to decide which of these is incorrect from the user's perspective. The only thing we can be clear about is that it probably is not a disagreement with respect to a selection rule.

Problem 2. For the second problem, where icons remain on the screen and become hollow when a disk is ejected, instead of disappearing, we could suggest that it is an operator mismatch. This is because the operator within the system was not what the user expected (of course, it might argued, at this point, that the terms operator and operation are being used as synonyms). Again, however, we could argue that this is a goal mismatch, as the goal stack the user and the system hold at this particular point are not identical, as one party (the user) has a goal to eject the disk and all its icons, and the other (the system) has a goal to eject the disk but retain some information regarding the disk.

Problem 3. In this 'disk swap problem', where users mistakenly believed that they had acted incorrectly in some way, we could characterize this as a goal mismatch because the subjects were unaware of the system goal. Once again, however, it is possible to suggest that this is a method or operator mismatch, as the goals were the same (to open the MacWrite application), but the expected methods were not agreed.

It is possible to argue that any mismatch is a goal mismatch or an operator mismatch, depending on the perspective we adopt. Moreover, if we decide that goals are agreed, then there is still a problem with the question of whether it is an operator or method that is at fault, given that a method is a sequence of operators and an operator, depending upon the grain of analysis, can be broken down into further operators. As a result, it seems as though a decompositional system such as GOMS, with goals and subgoals, and operators that can, on analysis, be broken into further operators and called a method, cannot work as a classification scheme.

Overall. We can say two things about this analysis. The most obvious is that a categorization of representational mismatches using GOMS is particularly difficult. It seems impossible to argue for a classification of any problem using this scheme that is not equivocal is some respect. The second point is that even if we agree a classification, deciding that a mismatch is a goal, operator or method mismatch does not seem to provide us with further useful information about the mismatch. With respect to errors then, the GOMS approximate theory appears to be of little use; it allows us no greater understanding of representational mismatches than ECM, and is more difficult to apply. Nevertheless, to be fair to Card, Moran and Newell, it is possibly worth mentioning that their production system theory was never intended for the analysis of errors.

Analyses: TAG

Payne and Green's (1986) Task-Action Grammar has constructs that are similar to ECM. For example, concepts and symbols are identified. TAG, however, goes further and distinguishes between simple tasks and rule schemata. During the analysis of the three problem descriptions mismatches will be assigned, not only to particular concepts and symbols, but also to these latter categories.

Problem 1. In the icons and windows problem (see earlier description) there appears to be a concept mismatch, as it is the concept of the relationship between the icon and its window that is unclear to the user. However, to take this classification further using the constructs provided by TAG, and hopefully gain a greater understanding of it, we can identify the problem as being with either a simple task or a rule schema. In this case we could suggest that it is a simple task, as the opening from an icon to a window is one of the systems basic operations, and does not involve the ordering, for the user, of a number of operations.

Problem 2. The second problem, with icons not disappearing when a disk was ejected, serves to demonstrate that TAG is likely to run into the same practical problems as ECM – that a problem can sometimes be classified in more than one way depending upon the perspective adopted by the classifier. In this case it is not clear as to whether it is a symbol or a concept that is mismatched (see earlier ECM analysis). If we decide, for the sake of argument, that it is a concept problem, then we could classify it as a problem with the rule schemata. This is because there are a number of simple tasks performed here. One simple task is selecting the disk to be ejected, and the other selecting the 'eject' command.

Alternatively, we might suggest that the problem is with the simple task, in that the simple task to eject a disk is not mutually agreed, in that the icons are left on the screen. Once again, we might retort that the simple task to eject a disk can be broken down into two simple tasks: one to eject the disk from the machine and the other to remove the icons (as this is possible if the disk icon is placed directly in the wastepaper basket icon.) Consequently, it is a problem with the rule schema in ordering the appropriate simple tasks. Of course, we could pursue this argument further, but it is doubtful whether much of use might be gained from it.

Problem 3. The final 'disk swap' problem can be characterized as a concept mismatch. To recapitulate, this is because the notion of the operation that is being performed, that of loading the data needed for the application in blocks from different disks, is not shared by the user and the system. With respect to TAG, the simple task of swapping a disk seems to be shared by both the user and the system. However, there is no understanding on the part of the user as to the need, on the part of the machine, for a large number of disk swaps. Consequently, we might argue that the rule schemata are mismatched. On the other hand, we might suggest that the notion of swapping disks is shared, but the need for repeating this operation is not understood, and so it is not a problem with either a simple task or a rule schema.

Overall. Given that it is possible to argue that some simple tasks can be decomposed into smaller simple tasks, TAG appears to be a decompositional scheme in the same way that GOMS is. In other words, whether a problem is classified as a simple task or rule schema mismatch depends upon the extent to which the problem has been broken down.

More importantly, however, is the question of does the process of classification using TAG really advance our understanding of representational mismatches. Ironically, thinking of slips and lapses in terms of simple tasks and rule schemata that organize this simple tasks has definite attractions: TAG's theory seems to better explain the errors that an expert might make (i.e. lapsing from one routine into another) rather than those that a user learning a system is likely to experience.

The applicability of HCI theories to errors

It might be possible to argue that the 'shoehorning' of problems into categories provided by theories that were never intended for this sort of use, does not show us a great deal that is useful. What is apparent from this exercise is that it is more difficult to classify the problems using TAG or GOMS than it is using the ECM scheme. One explanation of this could be that ECM has definitions specifically geared towards classification, whereas TAG and GOMS do not. Consequently, we might expect classification using ECM to be easier.

Another explanation is that both TAG and GOMS have problems in that their categories can be decomposed to a point where analysis is no longer useful. For example, goals can be divided up into subgoals and yet more subgoals, and operators divided up into smaller operators. Likewise, TAG's simple tasks can be divided up into smaller simple tasks. ECM does not appear to possess this problem. Although it is possible to decompose concepts into smaller concepts, etc., in reality this has not proved to be a problem.

Nevertheless, ECM still has the practical problem of multiple classification; where it is possible to classify a mismatch in more than one way depending upon the perspective that is adopted. In essence then, ECM is not afflicted by the decompositional problems evident when using GOMS or TAG. The most important point, however, is that while ECM is theoryless, TAG and GOMS are no more successful in providing a useful theoretical account of error.

The problems with classifying errors

Clearly TAG and GOMS have failed, as classification schemes, to provide anything more than ECM. Some may suggest that we require more sophisticated (and usable?) schemes. However, the problem may, at least in part, be with the approach of classification itself. Classification assigns a term to an error (e.g. knowledge-based, mistake, operation-symbol). Although such classifications may serve useful purposes, they are hardly explanations. For more useful explanations of errors we require more detailed theoretical accounts of how representations are formed, of how action sequences are created. Moreover, these accounts must explain why users sometimes act inappropriately (errors).

Part of the limitation with the classification approach has been elucidated by Hollnagel (1991), who uses a biological analogy to distinguish between the genotype (cause) of an error and the phenotype (manifestation) of an error. Some apparently quite different errors can have closely-related causes, while other identical error manifestations can have quite different causes. Yet a better understanding of these causes and manifestations cannot be provided by classification schemes alone. This point is further illustrated by Hollnagel's tongue-in-cheek (1989) suggestion of a Richter scale for errors – a measure he calls the Reason scale. From low to high the scale is as follows: an inaccuracy; a slip; an oversight; a mistake; a blunder; a violation. This scale is, of course, quite meaningless, and by placing this collection of categories in order of importance Hollnagel neatly illustrates the pointlessness of attempting to quantify actions (errors) without any fundamental understanding of their cause.

A meaningful Richter scale for errors would, of course, be of great value. The problem, however, is that to be useful classification schemes need to classify causes instead of manifestations. Classifying manifestations is relatively easy, but of limited use, while classifying causes is far more difficult. Moreover, without a general theoretical framework, and methods and tools for examining errors, such classification is almost impossible.

We have found that the common approach of classifying errors is limited, as we might expect, by the theory underpinning the classification. Classifying errors in itself provides no answer. Moreover, when we attempted to employ TAG and GOMS to gain a greater understanding of user error we failed. With hindsight we could claim that the failure of such (intentionally) limited theories was inevitable. Nevertheless, *our consideration of errors has provided an interesting demonstration of the evident need for broader theories in HCI.*

Cognitive science and HCI

To re-iterate, we require a theoretical underpinning for our analyses of errors, but where might this theory come from? A particular problem for the HCI field is that much of cognitive science is inaccessible to the designer. Few, if any, have the background or time to learn about the assumptions and approaches implicit within the areas that constitute cognitive science and psychology. As a result the limitations and potential uses of theories and ideas from this area are not understood. For many the terminology is vague and confusing, the concepts are difficult to fathom, and it is sometimes hard to distinguish a genuine advance from a change in fashion. One significant aspect of this problem might be best characterized as the *cognition club*, or, as some might refer to it, the *ahh, but not quite*, perspective. The latter refers to those ignorant of the theories and approaches within cognitive science checking that their understanding of a concept is correct, only to be told that there is yet another caveat.

The cognition club

One problem is that those in psychology, or the cognitive sciences if we prefer that term, know a great deal more than their theories suggest. We might consider the cognitive sciences as a club, although this is not a conspiracy, but the type of association that grows out of a shared culture. Indeed, some of the members may not even be aware of their membership. It nevertheless, acts as a club – a group of people with a private language. As a consequence it is acceptable to forward small but inadequate theories because all of the club members are aware of the sometimes unspoken caveats that accompany such theorizing. This tends to give rise to two sorts of theory: precise but inadequate, even if excellently written (e.g. Minsky 1975) or imprecise and rich – those papers that sometimes resemble a form of poetry, where so many invaluable and inexpressible ideas are given temporary form (e.g. Minsky 1981).

A further feature of this semi-exclusive club is that the members are quite willing, for the sake of convenience, to see the area divided up into quite artificial categories (i.e. thinking, memory, reasoning, decision-making, attention, etc.), because the membership knows that these areas cannot in reality be considered separate. Of course, the artificiality of this partitioning is rarely communicated to outsiders (although this is not intentional). Hence the theories, having so many unspoken caveats, are largely unusable. As Winograd (1981) comments, when dealing with memory, inference and attention, cognitive experimental psychology is not much use. Yet human–computer interaction requires something more precise, accurate and usable. Cognitive psychologists may know a great deal about cognition, but little of what they know is easily communicable.

Conclusion

We have seen how errors cannot presently be analysed in the depth that is required to best re-formulate design. The ECM scheme lacks a theoretical underpinning, and consequently produces a piecemeal analysis, without any necessary explanation of how users misunderstandings arise, and where to look if we wish to avoid them. The claims that the limited theories embodied within some of the analytic methods might make good this shortfall has not been supported. An alternative would be to return to cognitive psychology and science for a theoretical basis, Yet the theories within cognitive psychology and cognitive science are often subtle, limited to just a few aspects of cognition, and difficult to express in terms that can be easily understood by design team members.

It is suggested that HCI requires general framework theories, that not only explain the user's understanding, but how this understanding is then related to action in an interactive environment, so that action can be described, analysed and designed-for. Such a framework might be developed from current cognitive theories, and explain a broad range of phenomena, but should be presented in terms that can be easily communicated to designers who do not have the time to learn a great deal about psychology, cognitive science or artificial intelligence.

In short, more advanced techniques for analysing errors need to be built upon broad theories, rather than further classification schemes. Such methods need to allow designers to witness the falsity to their own beliefs regarding the user's actions and related understanding, within a general perspective of informing creative design, rather than prescribing it. Moreover, the focus must be upon all actions if an overall picture of the user's representation is to be developed, not just upon those actions judged to be in error.

References

Alty, J.L. and M.J. Coombs, 1980. University computing advisory services: The study of the man–computer interface. Software Practice and Experience 10, 919–934.

Bannon, L.J. and S. Bødker, 1989. Beyond the interface: Encountering artefacts in use. Report No. DAIMI PB – 288, Computer Science Department, Aarhus University, Denmark.

Barnard, P., J. Ellis and A. MacLean, 1989. 'Relating ideal and non-ideal verbalized knowledge to performance'. In: A. Sutcliff and L.A. Macaulay (eds.), People and computers V, Proceedings of the fifth conference on human–computer interaction. Cambridge: Cambridge University Press.

Barnard, P.J., N.V. Hammond, A. MacLean and J. Morton, 1982. Learning and remembering interactive commands in a text-editing task. Behaviour and Information Technology 1, 347–358.

Barnard, P.J. and M. Harrison, 1989. 'Integrating cognitive and system models in human–computer interaction'. In: A. Sutcliff and L.A. Macaulay (eds.), People and computers V. Proceedings of the fifth conference on human–computer interaction. Cambridge: Cambridge University Press.

Bellotti, V., 1988. 'Implications of current design practice for the use of HCI techniques'. In: D.M. Jones and R. Winder (eds.), People and computers IV. Proceedings of the fourth conference on human–computer interaction. Cambridge. Cambridge University Press.

Bjørn-Andersen, N., K.D. Eason and D. Robey, (eds.), 1987. Managing computer impact: An international study of management and organizations. Norwood, NJ: Ablex.

Booth, P.A., 1989. An introduction to human–computer interaction. Hove: Erlbaum.

Booth, P.A., 1990a. An explanatory framework for errors. Interacting with Computers, Noordwijkorhout, The Netherlands, 10th–11th December, 1990.

Booth, P.A., 1990b. 'ECM: A scheme for analysing user-system errors'. In: D. Diaper et al. (eds.), Human–computer interaction – Interact '90: Proceedings of the third IFIP conference on human–computer interaction, Cambridge. Amsterdam: North-Holland.

Booth, P.A., 1990c. Identifying and interpreting design errors. International Journal of Human–Computer Interaction 2, 307–332.

Booth, P.A., 1990d. Using errors to direct design. Knowledge-Based Systems 3, 67–76.

Booth, P.A., 1991. Modelling the user: User-system errors and predictive grammars'. In: G.A. Weir and J. Alty (eds.), HCI and complex systems. London: Academic Press.

Bowen, W., 1989. 'The puny payoff from office computers'. In: T. Forester (ed.), Computers in the human context: Information technology, productivity and people. Oxford: Blackwell.

Bradford, J.H., M.D. Murray and T.T. Carey, 1990. 'What kind of errors do Unix users make?'. In: D. Diaper, D. Gilmore, G. Cockton and B. Shackel (eds.), Human–computer interaction – Interact '90: Proceedings of the third IFIP Conference on human–computer interaction, Cambridge. Amsterdam: North-Holland.

Briggs, P., 1989. What we know and what we need to know: The user model versus the user's model in human–computer interaction. Behaviour and Information Technology 7, 431–442.

Brodbeck, F.C., D. Zapf, J. Prümper and M. Frese, 1989. Error handling in office work with computers: a field study. Unpublished manuscript, University of Munich, Department of Psychology.

Brooks, R., 1990. The contribution of practitioner case studies to human–computer interaction science. Interacting with Computers 2, 3–7.

Buckley, P.K. and J.B. Long, 1990. Using videotex for shopping – A qualitative analysis. Behaviour and Information Technology, 9, 47–61.

Card, S.K., T.P. Moran and A. Newell, 1983. The psychology of human–computer interaction. Hillsdale, NJ: Erlbaum.

Carroll, J.M. and C. Carrithers, 1984. Training wheels in a user interface. Communications of the ACM 27, 800–806.

Carroll, J.M. and M.B. Rosson, 1985. 'Usability specifications as a tool in iterative development'. In: H.R. Hartson (ed.), Advances in human–computer interaction. Norwood, NJ: Ablex.

Carroll, J.M. and J.C. Thomas, 1982. Metaphor and the cognitive representation of computer systems. IEEE Transactions on Systems, Man, and Cybernetics SMC-12, 107–116.

Clark, F., P. Drake, M. Kapp and P. Wong, 1984. 'User acceptance of information technology through prototyping'. In: B. Shackel (ed.), Human–computer interaction – Interact '84: Proceedings of the first IFIP conference on human–computer interaction, London. Amsterdam: North-Holland.

Cuomo, D.L. and J. Sharit, 1989. A study of human performance in computer-aided architectural design. International Journal of Human–Computer Interaction 1, 69–107.

Curtis, W., 1990. 'Empirical studies of programmers'. In: D. Gilmore, G. Cockton and B. Shackel (eds.), Human–computer interaction – Interact '90: Proceedings of the third IFIP conference on human–computer interaction, Cambridge. Amsterdam: North-Holland.

Davis, R., 1983a. User error or computer error? Observations on a statistics package. International Journal of Man–Machine Studies 19, 359–376.

Davis, R., 1983b). Task analysis and user errors: A methodology for assessing interactions. International Journal of Man–Machine Studies 19, 561–574.

Denning, P.J., D.E. Comer, D. Gries, M.C. Mulder, A. Tucker, A.J. Turner and P.R. Young, 1989. Computing as a discipline. Communications of the ACM 33, 9–23.

Détienne, F., 1990. 'Difficulties in designing with an object-oriented language'. In: D. Diaper, D. Gilmore, G. Cockton and B. Shackel (eds.), Human–computer interaction – Interact '90: Proceedings of the third IFIP conference on human–computer interaction, Cambridge. Amsterdam: North-Holland.

Doyle, J.R., 1990. Naïve users and the Lotus interface: A field study. Behaviour and Information Technology 9, 81–89.

Eason, K.D., 1976. Understanding the naive computer user. The Computer Journal 19, 3–7.

Eason, K.D., 1984. Towards the experimental study of usability. Behaviour and Information Technology 3, 133–143.

Eason, K.D., 1988. Information technology and organizational change. London: Taylor and Francis.

Frese, M., F.C. Brodbeck, D. Zapf and J. Prümper, 1990. 'The effects of task structure and social support on users' errors'. In: D. Diaper, D. Gilmore, G. Cockton and B. Shackel (eds.), Human–computer interaction – Interact '90: Proceedings of the third IFIP conference on human–computer interaction, Cambridge. Amsterdam: North-Holland.

Gaines, B.R., 1981. The technology of interaction – Dialogue programming rules. International Journal of Man–Machine Studies 14, 133–150.

Gavaghan, H., 1990. Human error in the air. New Scientist, 1743.

Gilmore, W.E., D.I. Gertman and H.S. Blackman, 1989. User–computer interface in process control: A human–factors engineering handbook. London: Academic Press.

Goodman, B.A., 1987. 'Repairing reference identification failures by relaxation'. In: R.G. Reilly (ed.), Communication failure in dialogue and discourse. Amsterdam: Elsevier.

Grant, S. and T. Mayes, 1991. 'Cognitive task analysis?' In: G. Weir and J. Alty (eds.), Human–computer interaction and complex systems. London: Academic Press.

Green, T.R.G., 1987. Limited theories as a framework for human–computer interaction. Invited address to the Austrian Computer Society's 6th Annual Interdisciplinary workshop. Mental models and human–computer interaction. Schårding, 9th–12th June, 1987.

Guindon, R., 1990. Designing the design process: Exploiting opportunistic thoughts. Human–Computer Interaction 5, 305–344.

Hackett, G., 1990. Sloan Management Review 31, 97–103.

Harding, C. and R.E. Rengger, 1989. The first steps towards a usability problem description language. Report DITC 154/89, National Physical Laboratory, Teddington.

Hiltz, S.R., 1980. 'Some results of the evaluation of the operational trials of the Electronic Information Exchange System'. In: A. Benenfeld and B. Kazlauskas (eds.), Communicating information: Proceedings of the 43rd ASIS Annual meeting 17. White Plains, NY: Knowledge Ind.

Hoerr, J., M. Pollock and D. Whiteside, 1989. 'Management discovers the human side of automation'. In: T. Forester (ed.), Computers in the human context: Information technology, productivity and people. Oxford: Blackwell.

Hollnagel, E., 1983. Human error. Position paper for NATO conference on human error, Bellagio, Italy, August.

Hollnagel, E., 1989. From human error to erroneous actions. Paper forwarded by author.

Hollnagel, E., 1991. 'The phenotype of erroneous actions'. In: G.A. Weir and J. Alty (eds.), HCI and complex systems. London: Academic Press.

Howes, A. and S.J. Payne, 1990. 'Supporting exploratory learning'. In: D. Diaper, D. Gilmore, G. Cockton and B. Shackel (eds.), Human–computer interaction – Interact '90: Proceedings of the third IFIP conference on human–computer interaction, Cambridge. Amsterdam: North-Holland.

Janosky, B., P.J. Smith and C. Hildreth, 1986. Online library catalog systems: An analysis of user errors. International Journal of Man–Machine Studies 25, 573–592.

Jones, S., 1984. Preliminary observations of Macintosh Plus users. Notes forwarded by the author, Department of Computer Science, Stirling University.

Kieras, D.E. and P.G. Polson, 1985. An approach to the formal analysis of user complexity. International Journal of Man–Machine Studies 22, 365–394.

Knowles, C., 1988. 'Can Cognitive Complexity Theory (CCT) produce an adequate measure of system usability?' In: D.M. Jones and R. Winder (eds.), People and computers IV. Proceedings of the fourth conference on human–computer interaction. Cambridge: Cambridge University Press.

Lewis, C. and D.A. Norman, 1986. 'Designing for error'. In: D.A. Norman and S.W. Draper (eds.), User centred system design: New perspectives on human–computer interaction. Hillsdale, NJ: Erlbaum.

Long, J.B. and J. Dowell, 1989. 'Conceptions of the discipline of HCI: Craft, applied science and engineering'. In: A. Sutcliff and L.A. Macaulay (eds.), People and computers V. Proceedings of the fifth conference on human–computer interaction. Cambridge: Cambridge University Press.

MacLean, A., P.J. Barnard and M.D. Wilson, 1985. 'Evaluating the human interface of a data entry system: User choice and performance measures yield different tradeoff functions'. In: P. Johnson and S. Cook (eds.), People and computers: Designing the interface. Proceedings of the first conference of the BCS HCI specialist group. Cambridge: Cambridge University Press.

Majchrzak, A. and D. Roitman, 1989. 'A socio-technical framework for integrating social and technical features of computer automated manufacturing'. In: M. Smith and G. Salvendy (eds.), Work with computers: Organizational, management, stress and health aspects: proceedings of the third international conference on human–computer interaction, Boston, MA. New York: Elsevier.

Marshall, C.J., B. McManus and A. Prail, 1990. Usability of product X – Lessons from a real product. Behaviour and Information Technology 9, 243–253.

Marshall, C.J., C. Nelson and M.M. Gardiner, 1987. 'Design guidelines'. In: M.M. Gardiner and B. Christie (eds.), Applying cognitive psychology to user–interface design. Chichester: Wiley.

McCosh, A.M., 1984. Factors common to the successful implementation of twelve decision-support systems and how they differ from three failures. Systems, Objective, Solutions 4, 17–28.

McMillan, T.C. and B.P. Moran, 1985, Command line structure and dynamic processing of abbreviations in dialogue management. Interfaces in Computing 3, 249–257.

Minsky, M.L., 1975. 'A framework for representing knowledge'. In: P.H. Winston (ed.), The psychology of computer vision. New York: McGraw-Hill.

Minsky, M., 1981. 'K-lines: A theory of memory'. In: D.A. Norman (ed.), Perspectives on cognitive science. Norwood, NJ: Ablex.

Monk, A.F., 1986. Mode errors: A user-centred analysis and some preventative measures using key-contingent sound. International Journal Man–Machine Studies 24, 313–327.

Moran, T.P., 1981. The command language grammar: A representation for the user interface of interactive computer systems. International Journal of Man–Machine Systems 15, 3–50.

Moran, T.P., 1986. Analytical performance models: A contribution to a panel discussion'. In: M. Mantei and P. Orbeton (eds.), Human factors in computer systems – III: Proceedings of the CHI '86 Conference, Boston, MA. Amsterdam: Elsevier.

Mowshowitz, A., 1976. The conquest of will: Information processing in human affairs. Reading, MA: Addison-Wesley.

Norman, D.A. 1983. Design rules based on analyses of human error. Communications of the ACM 26, 254–258.

Norman, D.A., 1988. The psychology of everyday things. New York: Basic Books.

Papert, S. 1980. Mindstorms: Children, computers and powerful ideas. New York: Basic Books.

Payne, S.J. 1984. 'Task-action grammars'. In: B. Shackel (ed.), Human–Computer Interaction – Interact '84: Proceedings of the first IFIP conference on human–computer interaction, London. Amsterdam: North-Holland.

Payne, S.J. and T.R.G. Green, 1986. Task-action grammars: A model of the mental representation of task languages. Human–Computer Interaction 2, 93–133.

Pomfrett, S.M., C.W. Olphert and K.D. Eason, 1984. 'Work organization implications of word-processing'. In: B. Shackel (ed.), Human–computer interaction – Interact '84: Proceedings of the first IFIP conference on human–computer interaction, London. Amsterdam: North-Holland.

Rasmussen, J., 1983. Skills, rules, knowledge; signals, signs and symbols; and other distinctions in human performance models. IEEE Transactions on Systems, Man and Cybernetics SMC-13(3).

Rasmussen, J., 1985. Trends in human reliability analysis. Ergonomics 28 1185–1195.

Rasmussen, J., 1986. Information processing and human–machine interaction. Amsterdam: North-Holland.

Reisner, P., 1981. Formal grammar and human factors design of an interactive graphics system. IEEE Transactions on Software Engineering SE7, 229–240.

Reisner, P., 1990. 'What is inconsistency?' In: D. Diaper, D. Gilmore, G. Cockton and B. Shackel (eds.), Human–computer interaction – Interact '90: Proceedings of the third IFIP conference on human–computer interaction, Cambridge. Amsterdam: North-Holland.

Riley, M. and C. O'Malley, 1984. 'Planning nets: A framework for analyzing user–computer interactions'. In: B. Shackel (ed.), Human–computer interaction – Interact '84: Proceedings of the first IFIP conference on human–computer interaction, London. Amsterdam: North-Holland.

Roberts, T.L. and T.P. Moran, 1983. The evaluation of text editors: Methodology and empirical results. Communications of the ACM 26, 265–283.

Rouhet, J.C. and M. Masson, 1990. 'Do pilots make special kinds of errors that we don't? The ARCHIMEDE research programme on civil aviation accidents involving human error'. In: P.C. Cacciabue and G. Mancini (eds.), Proceedings of the Ninth European Annual Conference on Human Decision-Making and Manual Control, Ispra, Italy, 10th–12th. September 1990. Commission of the European Communities, Joint Research Centre, Institute for Systems Engineering and Informatics.

Smith, S.L. and J.N. Mosier, 1984. Design guidelines for user–system interface software. Report ESD-TR-84-190, Mitre Corporation, Bedford, MA.

Thimbleby, H.W., 1982. Character-level ambiguity: Consequences for user interface design. International Journal of Man–Machine Studies 16, 211–225.

Thimbleby, H.W., 1990. User interface design. Wokingham: Addison-Wesley.

Welty, C., 1985. Correcting user errors in SQL. International Journal of Man–Machine Studies 22, 463–477.

Winograd, T., 1981. 'What does it mean to understand language?' In: D.A. Norman (ed.), Perspectives on cognitive science. Norwood, NJ: Ablex.

Wroe, B., 1986. 'Toward the successful design and implementation of computer-based management information systems in small companies'. In: M.D. Harrison and A.F. Monk (eds.), People and computers: Designing for usability. Proceedings of the Second Conference of the BCS HCI specialist group. Cambridge: Cambridge University Press.

Young, R.M. and A. Hull, 1982. 'Cognitive aspects of the selection of viewdata options by casual users'. In: Williams (ed.), Pathways to the information society. Proceedings of the 6th International Conference on Computer Communication, London. Amsterdam: North-Holland.

Acta Psychologica 78 (1991) 97–110
North-Holland

Knowledge retrieval and frequency maps *

Gill M. Brown

University of Manchester, Manchester, UK

This paper attempts to reconcile some of the explanations of expert error (i.e. slips and lapses) with some of the theory relating to the use of frequency information. Although there have previously been suggestions that novice errors may be accounted for in terms of frequency gambling, it has been assumed that experts do not use anything as crude as frequency. A study is reported that investigates the use of frequency information during learning in an everyday context. An· analysis of the errors that subjects made during this study, as they learnt more about the domain, suggests that 'experts' use frequency to as large a degree as novices. The factor that may differentiate experts from novices in any domain is not just the possession of domain knowledge, but also a sophisticated frequency map of the domain. Overall, these results suggest that expert errors are just as likely to be influenced by frequency of encounter in the everyday world as novice errors.

Introduction

What is Cognitive Ergonomics? Cognitive Ergonomics is concerned with the application of cognitive theory and the findings from Cognitive Psychology to real-world problems. One possible difficulty with this is that we know a great deal about how people behave in laboratories and particularly about what Kolodner and Riesbeck (1986; also Neisser 1978) call low knowledge experiments, for example dealing with unrelated lists of words – memorizing and recalling them, etc. However, it is difficult to see how much of this type of work can be applied to the real world, when we know that factors such as context are so important.

* Many thanks to Jim Reason, my supervisor, for his guidance and support. I am also grateful for the helpful comments on an earlier draft provided by Paul Booth and two anonymous reviewers. This work has been supported by SERC in the form of a postgraduate studentship.

Author's present address: G.M. Brown, Dept. of Clinical Psychology, University of Liverpool, Liverpool L69 3BX, UK.

Consequently, the work reported here has been approached with the aim of dealing with high-knowledge real-world domains. This research, concerned as it is with the use of knowledge in everyday contexts, involved the showing of a series of videos in the form of a story. Moreover, it provided an opportunity to monitor how people used frequency information as learning progressed. Although this type of stimuli may have a number of limitations, i.e. actually recording the learning process and the nature of the story itself, this method does have its advantages. Specifically, the material presented is controlled, and the objective content of different aspects of the videos (e.g the characters and the places) can be calculated with respect to subjects' exposure to these aspects. Therefore, this appears to be a useful and interesting way of looking at how people behave in a more dynamic everday learning situation.

Errors

Having outlined the approach adopted in this reseach, we need to discuss the central aim of this paper, which is to consider errors and how people behave in everyday contexts generally. Reason (1986, 1990; Reason and Mycielska 1982) has looked at this area in some depth, as has Norman (1981, 1983). They have provided detailed accounts of the kind of errors people make when they are very familiar with a task, such as lapses and slips, etc. Reason has also considered how people deal with the world and the role of frequency in cognition. More specifically, Reason has developed a model of knowledge retrieval: Cognitive Underspecification (1986, 1990) which is derived from the well-established observation that whenever cognitive operations are insufficiently specified, they tend to 'default' to contextually appropriate high frequency responses. This model identifies three distinct strategies of knowledge retrieval: automatic similarity matching between retrieval cues and stored knowledge attributes; frequency gambling to resolve 'fuzzy' matches (i.e. where a number of items have been partially matched, and where selection favours the most frequently encountered item); and inference, involving the largely conscious application of logical operators and selected knowledge items brought to mind by similarity-matching and frequency gambling. Furthermore, Reason suggests that although all of these strategies are likely to be involved in knowledge retrieval, their relative contributions are expected to vary: firstly, with the nature of the search task; and secondly, with the amount of domain knowledge or expertise possessed by the searcher.

Arising out of this is the suggestion that frequency determines errors in novices. Certainly, errors in answers to detailed questions about US States (Brown 1990) and US Presidents (Reason 1986) reveal a strong frequency bias. However, we know that experts also make errors of the kind that Reason and Norman have highlighted (i.e. those related to automaticity). But the question remains as to how experts make errors if they are not based on frequency? There are essentially two views concerning the use of frequency in the everyday world. The first, is that it is crude and is used less and less as we learn more about a domain. The second view is that its use may be crude at the beginning of learning, and novices may be over-reliant on it, but as we develop our knowledge of a particular domain we also develop a frequency map for that domain. In other words, our ideas about the relative frequencies of different items in different contexts become more finely tuned. Consequently, rather than using frequency less and less, we continue to use it but in a more sophisticated way, in a way that more carefully matches the particular domain in which we are working. It is these two possibilities that we are concerned with here. Frequency is important for novices, but is it as important to the experts who have a good understanding of a particular task, system, domain or machine?

In support of the view that frequency is important to both experts and novices, is the idea that frequency fits in very well with the the work that has been carried out from a connectionist perspective (also referred to as artificial neural networks or Parallel Distributed Processing; Rumelhart et al. 1986). According to this approach, frequency is represented by the strength of the links between various nodes and the patterns across the nodes within the networks/distributed system. However, although this may be an attractive argument it cannot be regarded as proof. The study reported below provides some support for the idea that frequency is used by experts.

A longitudinal study

The topic of study consisted of a series of six BBC videos of 'The Forsyte Saga' by John Galsworthy. The control group involved each of six different groups of subjects watching just one of the six videos, while the experimental group watched all six.

Rationale

The decision to select the Forsyte Saga as the course of instruction for this study was taken for a number of reasons. Firstly, one of the main considerations was the independent variable, i.e. frequency of encounter (FOE) or the number of times the subjects would be exposed to an entity or item, could be carefully controlled. Previous studies have been forced to use concepts already known to subjects, where subjects may have differed in their exposure to these concepts. Moreover, it was not clear, in these studies, to what extent subjects ratings of frequency reflected actual frequency in the everyday world. Secondly, the actual nature of the study material is not cumulative, unlike other 'taught' courses. For example, learning a foreign language usually involves, among other things, memorizing lists of verbs and their structure in different tenses, and then using them along with non-verbs in speech or written work. On the other hand, *The Forsyte Saga* represents a 'story', the nature of which prevents items from being learned in equal frequency – quite simply, some characters (items) occur more frequently than others. Moreover, this knowledge domain may be described as 'dynamic' when compared to other 'static' domains. In the former the story and characters change from video to video, while in the latter, individuals have a particular level of knowledge about that domain. Overall, the study provided an opportunity to look at human knowledge retrieval and the role of frequency in a controlled way that was, nevertheless, related to a story in an everyday context.

Hypotheses

(1) The first hypothesis deals with the assumption that subjective estimates of frequency are directly related to the representation of concepts and their frequencies within cognition, and that this representation is assumed to bear a direct relationship to frequency in everyday life. This study provided an opportunity to test this assumption, by correlating the subjective ratings of characters' frequency (provided by experimental group subjects) with char-

acters' objective frequency data (derived from their time 'on screen' in each video).

(2) The second hypothesis considers whether there is any frequency effect. In other words, when individuals in the experimental group respond incorrectly to the questions in the six recognition tasks is there a significant difference between the objective frequency for these incorrect answers compared to the correct frequency score?

(3) Assuming that there is a frequency effect we can now go on to consider the third hypothesis which is whether there is any difference in the frequency use of novices and experts. Firstly, there is the possibility that experts do not use frequency information greatly, on the other hand it may be the case that there is little or no difference and experts use frequency just as much as novices.

(4) Finally, it has been assumed that the two groups of subjects can be called 'experts' and novices because of the difference in knowledge between the two groups. I.e. the experts (experimental group) have seen all the videos, while the six control groups have watched one video each. (The term 'expert' is used loosely here to mean subjects with greater domain experience). For this assumption to be valid we must be able to establish that the experimental group's (experts') performance, in terms of their number of correct answers for each video, will exceed that of the control group (novices).

Method

Design

The study used a repeated measures design. All the 28 experimental group subjects took part in each of six conditions (i.e. watched all six videos over the six week period), whereas the 60 control group subjects (independent subjects design) watched just one video each (10 subjects per video). Order effects were not a consideration for the experimental group, as the videos represented a story and had to be shown in order.

Independent variables: Time/knowledge of characters,
Frequency of encounter for each character.
Dependent variables: Number answers correct,
Error frequency score.

The calculation of this last measure will be explained in the appropriate results subsection.

Subjects

Twenty-eight subjects took part in the study (experimental group). They were all first and second year psychology under-graduates studying at Manchester University. The only relevant criteria for selecting the subjects that volunteered to take part was that they had not previously seen any of the videos, read the books or watched the series on television. Twenty females and eight males took part in the study and their ages ranged from 18–22, with the exception of one subject aged 29.

The control group consisted of 60 subjects. Thirty-five were first and second year psychology undergraduates studying at Manchester University. The remaining 25 subjects were computing undergraduates studying at Manchester Polytechnic. Again, the only criterion for taking part in the study was that the subjects had not previously seen any of the Forsyte Saga videos, read the books or watched the series on television. There were 42 females and 18 males in the control group. There ages ranged from 18 to 32.

Materials

The recognition test was in the form of a multiple choice questionnaire, where a question was asked and the subjects were required to circle one of the six responses 'a' to 'f' (characters name). For example, one question was 'Who was referred to as "the most typical Forsyte of them all"', with the following six characters to choose from: Soames, Frances, Nicholas, Winifred, Swithin, Aunt Ann. Each of the 6 recognition tests consisted of 12 multiple choice questions. The control group subjects were given exactly the same recognition tests as for the experimental group.

Procedure

The investigation took place over a period of 6 weeks. The 28 subjects in the experimental group, were shown the six videos, one a week, in order. After each video, they were asked to complete the recognition test. The average running time for the videos was approximately 100 minutes each (1 hour 40 minutes) and the recognition task took approximately 15 minutes to complete. In contrast to this, for the control groups, a number of showings were timetabled and subjects were asked to come and watch just one video. After watching the video, they were asked to complete the accompanying recognition questionnaire.

Results

Do subjective measures of frequency correlate with objective measures?

In previous studies, regarding US presidents, US states, fruits and the like, the frequency of any particular concept has been assessed in a separate study by asking a

Table 1
Correlations between mean subjective estimates and mean objective measures (in seconds) of frequency over all six videos for the experimental group only.

	Mean	SD	N	St. Error	Skewness	Kutosis
Total: Subjective frequency	1.587	1.343	52	0.186	1.180	0.768
Total: Objective frequency	379.292	339.054	52	47.081	1.537	2.209

Pearson's product moment correlation: $df = 50$, $r = 0.90588$, $p < 0.0001$
Spearman's rank correlation coefficient: $df = 50$, $R = 0.7483$, $t = 7.976$, $p < 0.0001$

group of subjects to rate (on a scale of 1 to 7) the relative frequency of encounter (saliency) of each item, concept or character. The relative frequency for each concept is then computed as the mean average of all of the subjects' ratings. These averages are assumed to bear a direct relation to the representation of concepts and their frequencies within cognition, and this representation is assumed to bear a direct relationship to frequency in everyday life. This is a crucial assumption of all of these studies and, although such an assumption may not be unreasonable, the study presented here provides an opportunity to assess its validity. Each experimental group subject was asked to rate the frequency of each character in each Forsyte Saga video. These average ratings were compared with the objective time each character spent on camera in each video.

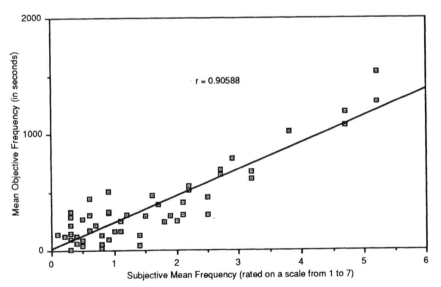

Fig. 1. A scattergraph showing subjective mean frequency for all six videos against objective mean frequencies for all six videos. The objective frequency (time on camera) was measured in seconds. The scale does not reach to 7 as none of the characters rated achieved an avergae rating above six.

Table 2
The frequency scores for correct and incorrect responses.

	Mean	SD	N
Incorrect answer frequency scores	21.844	14.377	657
Correct answer frequency scores	19.367	15.323	657
	$F(1/1312) = 9.126$, $p = 0.003$		

From the correlation in table 1 and the scattergraph in fig. 1 it is clear that the assumption regarding subjective measures of frequency was and is justified given the significance of the correlation between these measures and the objective ratings of frequency, based upon time on screen.

Does frequency of encounter affect knowledge retrieval?

The most obvious question that arises when considering this data, is *does frequency have an effect on knowledge retrieval?* If frequency of encounter has no effect then we might expect the average frequency of incorrect responses to be no higher than the average frequency for correct responses for the same questions. Consequently, if there is an effect, then we might expect a significant difference between the frequency score for the incorrect response and the frequency score for the correct response for each question incorrectly answered. In short, if frequency does play a significant role then the mean frequency for incorrect responses should be higher than the mean frequency for the correct responses.

As predicted, the frequency scores for the incorrect answers were higher than the frequency scores for the correct responses to the same questions (see table 2). Moreover, this result was highly significant ($p < 0.003$). The result suggests that the frequency of encounter of items in the world (or in this case in the videos that the subjects were shown) affects the way in which subjects retrieve their knowledge. That is to say, that when cognitive operations were underspecified during the questionnaires that were issued after each video, subjects tended to default to contextually appropriate high frequency responses.

Error frequency scores

Here, we consider the central question of whether subjects, as they learn more and become more familiar with the experimental stimuli, use frequency to a lesser degree. Alternatively, they may use frequency information to the same degree as novices. To answer this question we need to consider the error frequency scores for the experimental group as they progress through the videos. The mean error frequency difference was calculated using the following formula:

$$\frac{\Sigma\left(\frac{i-c}{n}\right)}{N},$$

Table 3
The difference in frequency scores for correct and incorrect responses (experimental group). The p value marked with an asterisk denote significant relationships in the opposite direction to that predicted.

Experimental Group	Mean	SD	N
Video 1: Incorrect freq. scores	28.789	10.894	160
Video 1: Correct freq. scores	17.866	8.897	160
$F(1/318) = 96.506, \ p < 0.0001$			
Video 2: Incorrect freq. scores	19.816	14.402	103
Video 2: Correct freq. scores	14.146	7.468	103
$F(1/204) = 12.581, \ p = 0.0008$			
Video 3: Incorrect freq. scores	20.895	17.322	86
Video 3: Correct freq. scores	31.953	19.011	86
$F(1/107) = 15.898, \ p = 0.0003$ *			
Video 4: Incorrect freq. scores	18.855	13.684	117
Video 4: Correct freq. scores	21.009	16.831	117
$F(1/232) = 1.154, \ p = 0.2837$			
Video 5: Incorrect freq. scores	18.380	12.029	108
Video 5: Correct freq. scores	22.231	18.120	108
$F(1/214) = 3.388, \ p = 0.0636$			
Video 6: Incorrect freq. scores	20.675	16.449	83
Video 6: Correct freq. scores	9.660	11.705	83
$F(1/164) = 24.706, \ p < 0.0001$			

where
i = the frequency score for the incorrect response,
c = the frequency score for the correct response,
n = the number of errors the subject made,
N = the number of subjects who committed errors.

The results for the experimental group can be seen in table 3. The results for the control group are provided for comparison purposes in table 4. The most important aspect of these results (explained in more detail elsewhere, Brown 1990), is that although the error frequency scores are in the opposite direction expected from some videos, the same trend is evident in the control group. This trend is evident in fig. 2.

Overall, we can conclude that there is no difference between the experimental and control groups in terms of their use of frequency. Therefore, it appears as though frequency is as important to experts as it is to novices. This, however, assumes that the

Table 4
The difference in frequency scores for correct and incorrect responses (control group). The p values marked with an asterisk denote significant relationships in the opposite direction to that predicted.

Control Group	Mean	SD	N
Video 1: Incorrect freq. scores	27.250	9.375	64
Video 1: Correct freq. scores	18.469	8.716	64
$F(1/126) = 30.117, \ p < 0.0001$			
Video 2: Incorrect freq. scores	20.293	13.704	58
Video 2: Correct freq. scores	16.621	8.352	58
$F(1/114) = 3.037, \ p = 0.0803$			
Video 3: Incorrect freq. scores	18.458	17.478	65
Video 3: Correct freq. scores	33.077	18.174	65
$F(1/128) = 21.848, \ p = 0.0001$ *			
Video 4: Incorrect freq. scores	22.897	16.010	58
Video 4: Correct freq. scores	20.241	16.001	58
$F(1/114) = 0.798, \ p = 0.3772$			
Video 5: Incorrect freq. scores	14.492	10.719	61
Video 5: Correct freq. scores	22.820	18.686	61
$F(1/120) = 9.117, \ p = 0.0034$ *			
Video 6: Incorrect freq. scores	17.474	14.757	78
Video 6: Correct freq. scores	11.615	12.762	78
$F(1/154) = 7.035, \ p = 0.0087$			

experimental group are experts by the time they have reached the end of the course of videos.

Number of correct responses

Although we might wish to conclude that frequency is as important to novices, how do we know that the experimental group are experts? In short, has learning occurred in the experimental group? Mean number correct is provided in table 5. A line graph of these results is provided in fig. 3.

From the results presented above it can be concluded that learning did occur in the experimental group, and the finding suggested in the previous results subsection, that frequency is as important to experts as it is to novices, is valid.

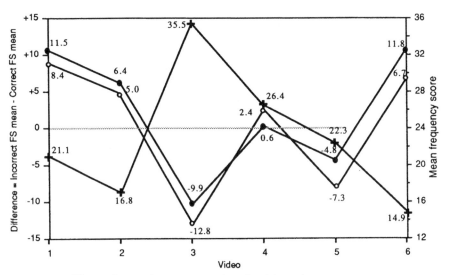

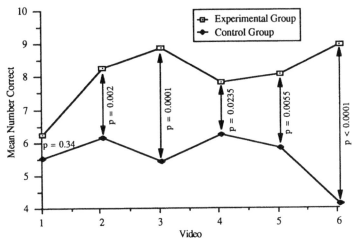

Fig. 2. A graph showing the difference between the mean frequency score for incorrect responses and the mean frequency score for correct responses, plotted against video number.

Fig. 3. A line graph showing the difference between the mean number correct for the experimental and control groups. The *p* values are for analyses of variance.

Table 5
The difference between the experimental group and control group for each video, in terms of the number of correct responses.

	Mean	SD	N
Video 1: Experimental group	6.286	1.941	28
Video 1: Control group	5.600	1.897	10
$F(1/36) = 0.930$, $p = 0.3434$			
Video 2: Experimental group	8.321	1.634	28
Video 2: Control group	6.200	1.874	10
$F(1/36) = 11.511$, $p = 0.0020$			
Video 3: Experimental group	8.929	1.844	28
Video 3: Control group	5.500	2.173	10
$F(1/36) = 23.208$, $p < 0.0001$			
Video 4: Experimental group	7.893	1.618	28
Video 4: Control group	6.300	2.406	10
$F(1/36) = 5.482$, $p = 0.0235$			
Video 5: Experimental group	8.143	2.085	28
Video 5: Control group	5.900	1.969	10
$F(1/36) = 8.760$, $p = 0.0055$			
Video 6: Experimental group	9.036	1.953	28
Video 6: Control group	4.200	1.476	10
$F(1/36) = 50.610$, $p < 0.0001$			

Discussion

There was no difference between the experimental and control group in terms of their use of frequency information. Nevertheless, despite the lack of any difference in the use of frequency between the two groups, the experimental group does appear to have undergone a learning process. Their performance exceeds that of the control group for all videos, except, of course, video 1. Similarity matching may have increased for experts, as the accuracy of these subjects increased with learning. However, there might be some doubt as to whether frequency gambling and similarity matching can really be separated, given the continuing use of frequency by experts. These concepts are metaphors and in reality they may well be part of the same process of knowledge retrieval. It is difficult to see how these processes might occur at different times – their strength is in operating together. Hence, the notion of a frequency map, based loosely on a neural net analogy, does provide a model for conceptualizing knowledge retrieval as a unified process. Moreover, this would explain why experts used frequency information to the same extent as novices.

The results of this study have shown that experts do use frequency information, but probably in a more refined way than novices – with a better developed frequency map. However, the underlying principle or basic premise of Underspecification – i.e. when cognitive operations are underspecified they tend to default to contextually appropriate high frequency responses, still holds. More importantly, it appears to be true not only for novices but for experts also.

To further relate these findings to everyday concepts, we can make a couple of speculations. Firstly, it appears as though learning about a domain involves acquiring a frequency map. We can speculate that this may be, at least in part why 'practice' is better than theory alone. In other words, individuals may experience how often things happen, and gain some idea of the likelihood of various things going wrong. Secondly, frequency information appears to degrade overtime (Brown 1990). A further speculation may be that becoming 'rusty' or 'out of practice' may involve loosing frequency information.

References

Brown, G.M., 1990. Knowledge retrieval and frequency maps. Ph.D. Thesis, Department of Psychology, Manchester University.

Kolodner, J.L. and C.K. Riesbeck 1986. 'Introduction'. In: J.L. Kolodner and C.K. Riesbeck (eds.), Experience, memory and reasoning. Hillsdale, NJ: Erlbaum.

Neisser, U., 1978. 'Memory: What are the important questions?' In: M.M. Gruneberg, P.E. Morris. and R.N. Sykes (eds.), Practical aspects of memory. London: Academic Press.

Norman, D.A., 1981. 'A psychologist views human processing: human errors and other phenomena suggest processing mechanisms'. In: Proceedings of the International Joint Conference on Articial Intelligence, Vancouver.

Norman, D.A., 1983. Design rules based on analyses of human error. Communications of the ACM 26, 254–258.

Reason, J.T., 1986. 'Congnitive underspecification: Its varieties and consequences'. In: B. Baars (ed.), The psychology of error. London: Plenum.

Reason, J.T., 1990. Human error. Cambridge: Cambridge University Press.

Reason, J.T., V. Horracks and S. Bailey, 1986. Multiple search processes in knowledge retrieval: Similarity matching, frequency gambling and inference. Unpublished report, Department of Psychology, Manchester University.

Reason, J. and K. Mycielska, 1982. Absent-minded? The psychology of mental lapses and everyday errors. Englewood Cliffs, NJ: Prentice-Hall.

Rumelhart, D.E., J.L. McClelland and the PDP Research Group, 1986. Parallel distributed processing: Explorations in the microstructure of cognition. Volume 1: Foundations. Cambridge, MA: MIT Press.

Acta Psychologica 78 (1991) 111–133
North-Holland

Event controllability in counterfactual thinking *

Vittorio Girotto

CNR, Rome, Italy

Paolo Legrenzi

Università di Trieste, Trieste, Italy

Antonio Rizzo

CNR, Rome, Italy

The counterfactual assessment of events, i.e. is the mental construction of alternatives to factual events, is a pervasive mental process that is quite natural for people. For example, people easily make counterfactual statements when reflecting on dramatic events ('If only I hadn't drunk alcohol the night of the car accident...'). The way in which people select the events to mutate when requested to undo a scenario outcome seems to be governed by general rules. One is that subjects tend to select exceptional (i.e. unusual or surprising) rather than normal events (Kahneman and Tversky 1982a,b; Kahneman and Miller 1986). Another is that subjects prefer to select the first rather than the subsequent events in a causal chain (Wells, Taylor and Turtle 1987). We hypothesized that events corresponding to controllable actions (i.e. voluntary decisions) by the protagonist of a scenario are more mentally mutable than events which occur in the surrounding background. In experiment 1, we manipulated the order and the controllability of four events in a scenario. Contrary to the causal order effect hypothesis, subjects preferred to change the event corresponding to a voluntary decision of the scenario actor, regardless of its relative position in the scenario. Experiment 2 showed that subjects made this choice regardless of the normal vs. exceptional status of the voluntary action event. Experiment 3 gave evidence that an unconstrained action performed by the focal actor of a story is more mutable than a constrained action performed by the same actor. The implications of these findings for the analysis of accidents involving human errors are discussed.

* This research was supported by grants from CNR Target Project 'Prevention and Control Disease Factor'; subproject 'Stress'; grant No. 9103615. We wish to thank Phil Johnson-Laird and Ann-Charlotte Welin for their helpful comments on an earlier version of the paper.

Requests for reprints should be sent either to V. Girotto or A. Rizzo, Istituto di Psicologia, CNR, Viale K. Marx 15, 00137, Rome, Italy.

Introduction

Counterfactual thinking, i.e. is the mental construction of alternatives to factual events, is a pervasive process that plays a significant role in several judgmental activities and emotional states, such as the assessment of causality and impact of actions and events (Dunning and Parpal 1989; Wells and Gavanski 1989), victim compensation (Miller and McFarland 1986), attribution of responsibility (Fabre, in press; Miller and Gunasegaram 1990), feeling of happiness or regret (Kahneman and Tversky 1982b; Johnson 1986; Landman 1987). (For a general review see Miller et al. 1990; cf. also Kahneman and Varey 1990). In general, the study of counterfactual thinking tries 'to identify the rules that determine which attributes of experience are immutable and which are allowed to vary in the construction of counterfactual alternatives to reality' (Kahneman and Miller 1986: 150).

In the analysis of event causation, counterfactual thinking has been proven to be of great relevance. Wells and Gavanski (1989) showed how an event to be considered causal must be psychologically mutable. For example, after a frontal crash between two trains running along a straight track, few people would consider the track among the causal factors. It is true that the nature of the track allows only to brake and does not permit to avoid an obstacle laterally. But tracks are an immutable and distinctive characteristic of trains. They are not considered as a possible 'cause'. Let's instead consider a frontal crash between two cars running along a road so narrow as to permit the passing of only one car. In this case, contrary to the case of the track, it is much more probable that the road would be considered a possible causal factor. Unlike this example, where two similar stories with different event mutability were confronted (track vs. road), Wells and Gavanski manipulated as variable the availability of alternatives to a given event of a story. In one of their experiments '...subjects read one of two versions of a story concerning a young woman, Karen, who went to a restaurant with her boss, Mr. Carlson. Mr. Carlson, unaware that Karen was allergic to wine, ordered a dish for her that contained a wine sauce. Kate ate the dish and had a severe allergic reaction from which she died. Mr. Carlson's behavior and the outcome were the same in both conditions; that is, he ordered the dish and Karen died. The only difference was that in the one story Mr. Carlson first considered ordering an alternative dish for Karen that did not contain wine sauce, whereas in the other story the alternative dish he considered also contained wine sauce' (Wells and Gavanski 1989: 163). Subjects attributed greater causal significance to the event which had a highly available counterfactual mutation that might have undone the dramatic outcome. That is, subjects considered more causal the choice made by Mr. Carlson in the story where he *considered* the dish that did not contain the wine sauce. This difference is interesting because, in both conditions, the action judged as more or less causal (Mr. Carlson's choice of the dish) was the same, it produced the same consequences (Karen's death), and these consequences were unpredictable to Mr. Carlson. What makes the difference is the fact that in one case we can more easily imagine an alternative choice eliminating the dramatic outcome, than in the other one.

According to Wells and Gavanski, the causal attribution theory (e.g. Kelly 1967) cannot make predictions concerning situations of this kind. However, the two models are not considered as conflictual. The principal distinction concerns the fact that traditional attribution theory restricts its analysis to given facts, whereas mental simulation involves a comparison between reality and what might have been (cf. also Lipe 1991).

The distinction between reality and counterfactual alternatives, has been shown to be of relevance also in the analysis of human errors in industrial accidents. For example, the INRS method (Institut National de Recherche et de Securité, cf. Leplat 1987) is based on an accident analysis that first considers the deleterious outcome, then draws a backward chart where the events are reported that, with their occurrence or missing occurrence, might have avoided the outcome. The events reported in the chart are by definition 'changes or variations in the *usual* conditions'. The behavior of the system is assessed precisely from the point of view of 'what might have been'.

Given the relevant role of counterfactual thinking in accident analysis and causal attribution, it is important to understand the cognitive processes that govern it, not only for theoretical reasons but also for possible ergonomic implications.

Mutability factors

So far, little is known about the way in which people generate alternative scenarios in real-world situations. However, laboratory research has indicated some general rules that govern the mental mutability of events.

The exceptionality of an event was the first mutability factor pointed out in the literature. As shown by Kahneman and Tversky (1982a), subjects who had to undo the outcome of a story tended to select *exceptional* (i.e. unusual or surprising) *more than normal events*. Kahneman and Tversky provided their subjects with a version of a story in which Mr. Jones was killed in a traffic accident when driving home after work. In one version, Mr Jones left work earlier than he usually did, but he took his usual route. In the other version, Mr. Jones left work at his usual time but took a different route home. Kahneman and Tversky asked their subjects to undo the outcome by finishing a sentence that began 'If only...'. Even though the outcomes in the two versions were the same and they did not differ in terms of their probability, subjects produced different and specific changes in the two cases. In the first version, most of them modified the time variable ('If only he had left work at his usual time...'). In the second one, most of the subjects modified the route ('If only he had taken his usual route...'). As Kahneman and Tversky concluded, subjects undid 'the accident by restoring a normal value of a variable [rather] than by introducing an exception' (1982a: 205).

According to Wells et al. (1987) a major factor governing the mental simulation of alternative scenarios is that causes are more mutable than effects. As a consequence, subjects should prefer to mutate the first rather than the subsequent events in a causal chain. Wells et al. presented their subjects with a scenario about a young man (William) who was going to a store in order to buy a stereo system on sale. His progress was impeded by four minor misfortunes: a speeding ticket, a flat tire, a traffic jam and a group of senior citizens crossing the street. William arrived at the store 35 min after the sale started only to find that the last stereo system had been sold just a few minutes before.

Subjects read one of four versions of this story. Each version had a different ordering of the event sequences, and the four events were presented in each of the four possible positions in the scenario. Subjects were asked to list the different ways in which the events in the story could be changed so that the outcome of the story would be different. It should be noted that the events of the story were linked in a causal way and that the modification of each event was sufficient to undo the outcome. Despite this formal equivalence, the results showed

that subjects tended to change the first event and showed no preference for the other events, regardless of their nature.

Following Wells et al. (1987) the two mentioned factors governing the mutability of an event can be unified in a single principle: On the one end, the first event in a causal chain is more mutable than the subsequent event(s) because it is perceived as not constrained by previous event(s) (at least events of the story). On the other end, normal events are less mutable than exceptional ones, because the former are perceived as constrained by previous causes (e.g. social and legal rules, habit and so on), and the latter are, by definition, events which occur in spite of rather than because of these constraints. Thus, in both cases, the mutability of an event appears to be inversely related to the number of causal conditions which constrain its occurrence.

This general principle is a plausible candidate for explaining the results of the previous literature. However, previous literature did not consider a distinction between two general classes of events of the real world, which might affect the generation of the counterfactual alternatives. This distinction concerns the events which correspond to the *actions* performed by an individual, and the *events which occur in the surrounding background*. Results reported by Turnbull (1981) showed that this distinction is present in the naive ontology of the subjects, and that it determines a number of interesting effects, including, in some cases, a feeling of potential control over chance.

If considered from the common sense point of view (which involves also the design of many industrial safety programs, see for instance Wagenaar et al. 1990), the actions performed by an individual are usually assumed to be independent from external causes (Sappington 1990). Therefore, individual actions are perceived as less constrained than the events which occur in the surrounding background. Now, if event mutability is a function of the number of perceived constraints, then the *events which correspond to individual actions should be more modifiable than the surrounding events* as far as these actions are *controllable*, that is, perceived as expression of free will. A theoretical discussion on the foundations of free will (see Dennett 1984; Dupuy 1990) is beyond the scope of this paper. For the present purposes, we define as controllable an event the occurrence of which depends on an actor's decision.

In the following three experiments, the role of the controllability factor in determining the construction of counterfactual alternatives has been studied in relation to the factors investigated in the previous literature. Its relevance for the analysis of system malfunction involving human action will be discussed in the Conclusions.

Experiment 1: Controllability and causal order

In experiment 1 the Controllability factor was compared with the Causal Order factor. As stated by the Causal Order hypothesis (Wells et al. 1987), subjects when requested to modify the outcome of a causal sequence of events, will tend to modify the first event in the sequence whatever its nature. Even though the influence of the causal order is not questioned, according to our hypothesis an event under the control of the protagonist of the story should be highly mutable independently from its position in the causal sequence. In the study by Wells et al. there is some empirical evidence that seems to corroborate the controllability hypothesis. In their first experiment, the only event that was partially controllable by the actor of the story (getting a speeding ticket) was significantly more mutated than the other events, which did not differ from each other.

In order to test this hypothesis, we created a story in which a causal chain ended with a tragic outcome. The story presented four principal events that could be balanced according to their order in the causal scenario. Three of the four events were independent of the will of the protagonist of the story, while the fourth was a controllable action. Following our hypothesis, subjects should prefer to mutate this latter event more than the others when requested to undo the tragic outcome.

Method

Subjects

The sample was composed of 108 male and female students of the introductory psychology and humanities courses. They were approached in a university cafeteria, and randomly assigned to one of four groups of equal size ($n = 27$).

Materials

Four versions of a scenario with a dramatic outcome were constructed. The story was about a bank employee, Mr. Bianchi, who worked in an agency situated in a village near to the one where he lived with his wife. The day of the accident he was going home after work but his progress toward home was delayed by three minor misfortunes and by an intentional decision: the manoeuvres of a lorry, the passage of a flock of sheep, an impeding tree trunk, and his own decision of entering a bar to drink a beer. When he arrived home, Mr. Bianchi found his wife on the floor. He realized that she had had a heart attack and she was dying. He tried to help her, but his efforts were in vain.

Each version of the story had a different ordering of the event sequences, and the four events were presented in each of the four possible positions in the scenario. Each scenario presented a causal relationship between events so that each event occasioned subsequent event(s). Yet the removal of each single event was sufficient for imagining a different outcome. For example, one version of the story presented the following sequence of events: (a) a lorry manoeuvring in the square where Mr. Bianchi was parked; (b) a flock of sheep crossing the road; (c) Mr. Bianchi's decision to drink a beer in a bar that was reopening for the evening; (d) a tree trunk lost by a tractor along the road. It should be noted that when the bar event occurred in all but the first position it was presented as having been occasioned by the previous delay (i.e. 'Mr. Bianchi arrived in the village just when the bar was opening for the afternoon break').

Design

   The design of the experiment was 4 (Order) × 4 (Event) in a conjugate Latin square 4 × 4 design.

Procedure

   The experimental procedure was similar to that used by Wells et al. (1987: exp. 1). Each participant was given a booklet containing one of four versions of the story and an answer frame. The booklets were randomly assigned to subjects. They had to read the story and to list, in order of importance, four different ways in which the events in the story could be changed so that the outcome would be different. They were informed that they could refer back to the story if they wished.

Results and Discussion

   The subjects' responses were tabulated by three independent coders. The intercoder reliability in categorization was higher than 90%.

   Disagreements were resolved via discussion. Six categories of modifications were identified: Bar, Lorry, Trunk, Flock, Others and Comments. The last category involved responses that did not change any element of the story but were comments on it (e.g., 'You cannot fight against fate'). Due to the low absolute frequencies (6 out of 416 responses) and its position order in the patterns of response (the fourth for any of them), this category was not further considered. The category Others involved mutations such as: [If only...] 'Mr. Bianchi had left work earlier'; 'Mr. Bianchi had lived in the same city where he worked'; 'Mr. Bianchi had phoned his wife (from the office, from the bar, along the road)'; 'there had been some neighbors'; 'the road had been larger'; 'Mrs. Bianchi had not been sick'; and so on. Since among these events there was no distinctive event, these events were not considered for the computation of the mutability indexes. For this reason, 6 subjects, whose mutations were all classified as 'Others', were discarded from the final analysis. In addition, 3 further subjects were discarded from the final analysis because in their answers it was not possible to identify the order of the mutations.

   The categories, frequencies and percentages of responses are reported in table 1.

   As can be seen in table 1, the four manipulated events cover 72.67% of all the changes produced by the subjects to modify the outcome of the story. For the experimental events two indexes of mutability were computed as in the study by Wells

Table 1
Experiment 1: Frequencies and percentages of the modified events.

Modified event	Frequency of response	% of all responses
Bar	97	25.0
Trunk	73	18.8
Flock	57	14.7
Lorry	55	14.2
Other	106	27.3

et al. (1987). The first index concerns the absolute frequency of subjects for whom a given event had been mentioned first in the mutations. This index was computed considering only the four manipulated events, excluding from the computation any mention of the category 'Others'. Table 2 shows this index of mutability of events in percentages as a function of the event position in the four scenarios.

A 4 (Event) $\times$ 4 (Order) contingency table analysis on the observed frequencies revealed a main effect both for Event, $\chi^2(3, N = 99) = 98.77$, $p < 0.001$; and Order, χ^2

Table 2
Experiment 1: Mean rank of event changes and percentage of first event changes, as functions of the position of the event in the scenario.

Event	Position				M across position
	1	2	3	4	
1. Bar					
Rank	1.56	1.96	2.29	2.04	1.97
%	87.0	57.7	70.8	57.7	68.3
2. Trunk					
Rank	2.77	2.92	3.91	3.54	3.14
%	38.50	15.40	4.30	8.30	16.60
3. Flock					
Rank	3.42	4.29	3.88	3.78	3.84
%	23.10	8.30	0.0	4.30	8.90
4. Lorry					
Rank	3.71	3.35	3.85	4.01	3.76
%	12.50	4.40	3.80	3.80	6.10
M over events					
Rank	2.89	3.13	3.35	3.36	
%	40.20	21.40	19.80	18.50	

Note: The rank refers to the order of position with which an event was reported in the ordered answer of the subjects (i.e., rank 2 attributed to the event X means that the event X was reported as the second mutation out of four). Only the experimental events were taken into account in the calculation of the rank.

(3, $N = 99$) = 11,16, $p < 0.025$. It was not possible to calculate the interaction between Event and Order because more than 20% of the frequency cells reported a frequency lower than 5. Neither was it possible to collapse the Order or Events factors because both the grouping of positions 2, 3 and 4 with respect to each single event, and the grouping of the events Lorry, Trunk and Flock with respect to each single position gave a frequency distribution where the percentage of cells with frequencies less than 5 was still higher than 20%. Two further analyses compared the first position vs. the remaining, and the Bar event vs. the other three events. Both analyses showed a significant difference: First position vs. other positions, $\chi^2(1, N = 99) = 4.04$, $p < 0.05$ (Yates' correction); Bar event vs. other events, $\chi^2(1, N = 99) = 11.68$, $p < 0.001$ (Yates' correction). The comparison between the three remaining positions showed no significant differences, while the comparison between Trunk, Flock and Lorry turned out significant: $\chi^2(2, N = 32) = 6.04$, $p < 0.05$ (see fig. 1a).

The second index of mutability of an event is marked by the rank representing the order in which the changes of the story are reported by the subjects. Since we had four ordinate changes as subject, rank 1 was assigned to the first experimental event mentioned, rank 2 to the second, and so on until rank 5 was assigned to any event that was not mentioned in the mutations proposed by the subject. In table 2 the mean rank of each event is reported as a function of the order in which they occurred in the scenario.

The data were analyzed by a two-way ANOVA, Event (4) × Order (4). Both the main effects turned out to be significant (Event, $F(3, 380) = 46.26$, $p < 0.001$; Order, $F(3, 389) = 3.37$, $p < 0.05$) but not their interaction. A post-hoc analysis (Fischer test, $p < 0.01$ for all comparisons) for the Event factor revealed that the Bar event had a higher rank than the other three events, and the Trunk event had a higher rank than the Lorry and Flock events. For the Order factor, a post-hoc analysis (Fischer test) indicated that the events presented in the first position were significantly ($p < 0.05$) more mutated than the events presented in the third and fourth position. Finally, a comparison between the four different scenarios showed that they did not differ from each other (see fig. 1b).

On the whole, these results show that, for both the mutability indexes, an event controllable by the protagonist of a story is more mutated than any event that occurs in the surrounding background. Indeed, we found that the Bar event is always the more mutated event, quite apart from its position in the causal scenario. It is worthwhile to note that, although the Order factor was significant (in accordance with the data obtained by Wells et al. 1987), the controllable event turned out to be more mutable than the other events, even when it was not presented in the first position. In other words, also in cases where the controllable event followed one or more events, of which it could be considered an effect, its mutability was higher than that of the event(s) that occasioned it.

Indirect evidence supporting our hypothesis comes from the qualitative analysis of changes proposed by the subjects for each event. We classified any single mutation as Active or Passive as a function of the role played by the protagonist in the counterfactual alternative of the story. For example, we considered Active mutations like the following: [If only...] '...Mr. Bianchi had not decided to drink the beer...'; '...Mr. Bianchi had helped in removing the trunk...', and we considered Passive mutations

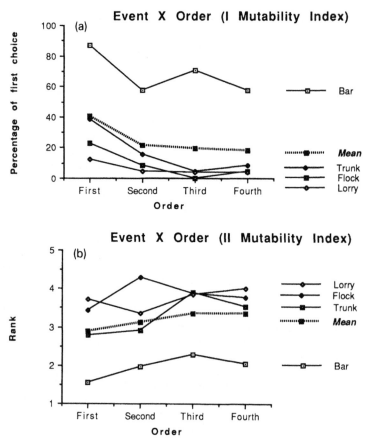

Fig. 1. Mutability of events as a function of their order in the scenario for the first (a) and the second (b) index.

like: [If only...] '...the bar had been closed...'; '...the trunk had not fallen...'. Although for the Bar event the mutations could have occurred in both directions, most changes (about 70%) were of the active kind (of the remaining, 23% were classified as passive, and 7% were not classified). For the other manipulated events of the story, even if these were out of control for the protagonist, in about 30% of the changes the subjects imagined Mr. Bianchi acting actively.

Experiment 2: Controllability and normality

As reported above, Kahneman and Tversky (1982a) produced experimental evidence supporting the hypothesis that exceptional events are more mutable than normal ones, and that in undoing the results of causal sequences people prefer to alter exceptional rather than routine aspects. In the stories by Kahneman and Tversky (1982a) the two levels of the Exceptionality factor were both actions performed by the protagonist: 'route toward home', and 'time of departure from the office'. As reported in one version of the story an event was described as habitual and the other as exceptional, in the other version their relationship was inverted. Therefore, the sequence of events producing the accident was different in the two cases. In other words, each candidate to the mutation should compete with partially different competitors in the two versions of the story. One can wonder if in two *identical* scenarios where a given action of the protagonist is presented in one case as normal and in the other as exceptional, the Exceptionality factor would influence the mutability of the action.

A controllable action, as based on a decision, is a choice between two or more competing behaviors, and thus, directly suggesting counterfactual alternatives. Defining such an action as exceptional should increase its mutability because the exceptionality specifies at least one potential counterfactual alternative, that is, the normal behavior (default alternative, Wells and Gavanski 1989). This being the case, two identical scenarios, except for the exceptionality/normality of a given action of the protagonist, should bring about a distribution of mutability of events where: (a) the controllable action would be more mutable than surrounding events; (b) the mutability of the action would increase when the action is exceptional, reducing the mutability of surrounding events.

We tested these predictions by using the same paradigm as in experiment 1. We constructed two identical versions of Mr. Bianchi's story. The only difference was that, in one case, the decision of the protagonist (entering the bar to drink a beer) was presented as habitual, in the other case, the decision was presented as exceptional. Given the difficulty to define an absolute degree of exceptionality for events (cf. Well et al. 1987), the exceptionality of the protagonist's decision was established on the basis of an intrapersonal norm ('Mr. Bianchi, as usual, decided to enter in the Bar...' *vs.* 'Mr. Bianchi, exceptionally for his habits, decided to enter in the Bar...').

Method

Subjects

The sample was composed of 90 students of a psychology course who participated on a voluntary basis.

Material

Two versions of Mr. Bianchi's story were constructed. In one version the decision to stop at the bar was described as usual, in the other version it was presented as unusual. In both conditions, the sequence of events was the following: (a) a flock of sheep crossing the road; (b) a tree trunk lost by a tractor along the road; (c) a lorry manoeuvring at the entrance of Mr. Bianchi's village; (d) Mr. Bianchi's decision to stop at the bar to drink a beer.

Table 3
Experiment 2: Frequencies and percentages of the modified events.

Modified event	Frequency of response	% of all responses
Bar	79	23.4
Trunk	68	20.2
Flock	58	17.2
Lorry	55	16.3
Other	77	22.9

Design and Procedure

A 2 (normal vs. exceptional Scenario) × 4 (Event) factorial design was adopted. The subjects were randomly assigned to one of the two conditions ($n = 45$), and individually tested. The procedure was the same as that used in the previous experiment.

Results and Discussion

The four experimental events covered 77.1% of the changes produced by the subjects to alter the outcome of the story (see table 3). For these events, the two mutability indexes used in experiment 1 were calculated (see table 4). Four subjects whose mutations were all classified as 'Others', were discarded from the final analysis.

In regard to the first mutability index, data were analyzed by a Contingency Table 2 (Condition) × 4 (Event). The comparison between the two conditions did not show a significant difference ($\chi^2(3, N = 86) = 2.45$). By contrast, in each condition, the comparison between events turned out to be significant (Normal condition: $\chi^2(3, N = 42) = 38,15, p < 0.001$; Exceptional condition: $\chi^2(3, N = 44) = 51.43, p < 0.001$). An analysis comparing the data of the Bar event vs. the data of the other three collapsed events, indicated a significant difference in the Exceptional condition ($\chi^2(1,$

Table 4
Experiment 2: Mean rank of event changes (II mutability index), and percentage of first event changes (I mutability index), in the two conditions.

Condition	Event			
	Flock	Trunk	Lorry	Bar
Normal				
Rank	2.80	2.59	3.57	2.0
%	11.9	21.4	4.8	61.9
Exceptional				
Rank	2.95	2.82	3.73	1.82
%	11.4	18.2	0	70.4

$N = 44) = 6.9$, $p < 0.01$, Yates' correction), but not in the Normal condition ($\chi^2(1,$ $N = 42) = 1.93$).

A two-way ANOVA, Condition (2) × Events (4), on the second index of mutability (rank scores), revealed a main effect for events ($F(3, 336) = 24.43$, $p < 0001$). But neither the condition factor ($F(1, 336) = 0.185$), nor the interaction Condition × Event were significant ($F(3, 336) = 0.376$). A post hoc analysis (Fisher test) indicated that the Bar event was significantly more mutated than the other three events, and that the Lorry event was significantly less mutated than the Trunk and the Flock events (for all comparisons, $p < 0.01$).

As for experiment 1, we classified each proposed mutation in terms of the active or passive role that the protagonist played in it. The response patterns were similar to those obtained in the previous experiment. The majority of mutations involving the Bar episode (65%) implied the active participation of the protagonist. Even for the external events, often the proposed mutation entails some potential actions of Mr. Bianchi (44%). These percentages were approximately the same in the two conditions.

Based on our hypothesis, we had predicted a significant main effect for the Event factor, and a significant interaction between the Event and the Condition factors. The obtained results revealed a significant main effect for the predicted factor, but not a significant interaction between the two factors. For both indexes of mutability, our subjects showed a significant preference for mutating the protagonist's action, regardless of its normal or exceptional status. Yet, the analysis of the scores of the first index of mutability indicated that the mutations of an *exceptional* controllable action are significantly more frequent than *all* the mutations of the other non-controllable events when collapsed in one category. On the contrary, the mutations of a normal controllable event did not differ significantly from the number of mutations of the other events collapsed in one category. However, apart from this effect, the present results indicate that a controllable action easily evokes counterfactual alternatives even when it is normal, that is, when a specific alternative is not immediately available (as is it true of an exceptional action, whose default alternative, i.e., the normal action, is highly available).

Experiment 3: Controllability and focus

It might be argued that the results of our experiments are not due to the controllable nature of the action performed by the protagonist, but to the differential mutability of focal and background events. Experimental evidence exists that the attributes of a focal actor are more mutable than those of non-focal ones. For example, Read (1985, quoted by Kahneman and Miller 1986) asked her subjects to indicate how a collision between the car of the protagonist of a story (Helen) and another car was avoided at the last minute. Although a critical action might be attributed to *both* car drivers, most of the subjects answered that Helen did something which prevented the accident. Now, in our scenarios, the controllable action performed by the protagonist (deciding to enter the bar) is the *most focal event*, i.e., compared to the other events, it is the event that concerns primarily the protagonist focus. Therefore, one might conclude that its high mutability depends on its focal nature rather than on its controllable nature. In order to rule out this interpretation, we presented subjects with a version of Mr. Bianchi's story in which all the episodes were actions performed by the protagonist, i.e., focal actions, but having a different degree of controllability. In other words, in this new version of the story, only one action was fully controllable (deciding to enter the bar). Two other actions were constrained by the protagonist's physical state (deciding to go back to the office in order to take a pair of glasses because the first pair broke, and deciding to stop the car in order to cure himself because of an asthma attack). If the primary factor in determining the mental mutability is the focal nature of events, then no difference in the choice of the three focal actions should appear. On the other hand, if the mental mutability is governed by the perceived degree of control of the protagonist over the scenario events, then the controllable action should be the first event changed by the subjects. Experiment 3 addresses this issue.

Method

Subjects

The sample was composed of 48 male and female introductory psychology students who participated on a voluntary basis.

Materials

A new version of Mr. Bianchi's story was constructed. Mr. Bianchi, when opening the door of his car parked near the bank, broke his glasses. This fact obliged him to go back to his office to look for an old pair of glasses that he remembered keeping there. Along the road, he stopped at a Bar to drink a beer. Finally, just before reaching his village, Mr Bianchi had an asthma attack. He stopped the car by the road to use an aerosol. The initial and final parts of the story were the same as in experiments 1 and 2. There was no causal relationship among the three events and only one ordering sequence.

Design and Procedure

In this experiment there was only one independent variable with three levels (Events). The procedure was similar to that used in the previous experiments.

Results and Discussion

As can be seen in table 5, the three experimentally manipulated events constituted the majority (85%) of the mutations produced by the subjects. For these events, the two

Table 5
Experiment 3: Frequencies and percentages of the modified events.

Modified event	Frequency of response	% of all responses
Bar	43	31.8
Glasses	39	28.9
Asthma	33	24.4
Other	20	14.8

Table 6
Experiment 3: Mean rank of event changes (II mutability index) percentage and frequencies of first event changes (I mutability index).

	Event		
	Bar	Glasses	Asthma
Rank	1.47	2.34	2.87
%	75.6	10.6	12.8
N	36	5	6

mutability indexes used in the previous experiments were calculated (see table 6). One subject whose mutations were all classified as 'Others' was discarded from the final analysis.

For both indexes the comparison between the three events turned out to be significant. The frequencies of the first event changes were significantly different ($\chi^2(2, N = 47) = 39.87$, $p < 0.001$). The great majority of the subjects (76%) mutated the Bar event first. A one-way ANOVA on the rank scores revealed a significant main effect ($F(2, 140) = 26.4$, $p < 0.001$). A post-hoc analysis (Fisher test, $p < 0.01$, for all comparisons) indicated that the subjects mutated the Bar event more readily than the Asthma and Glasses events. In addition, the latter event was significantly more mutated than the Asthma event. Finally, all except one (96%) of the mutations for the Bar event were classified as active. The mutations of the remaining events involved in a third of the cases (33%) the active participation of Mr. Bianchi. (These active mutations were more frequent for the Glasses event than for the Asthma event, 38% and 27%, respectively.)

These results supported the hypothesis that the mental mutability of an individual's action depends on the perceived constraints on its accomplishment. Despite the common reference to the protagonist of the story, subjects preferred to change the event which corresponded to a free decision, rather than events constrained by some causes external to the individual's intention, such as his body state. The higher mutability of the Glasses event over the Asthma event cannot be considered to be the result of an order effect, as the Glasses event, although presented before, was not causally linked with the Asthma event. The obtained difference suggests that there is a continuum in the degree of event controllability. Although both events were determined by problems concerning the body state of the protagonist, the asthma attack was a more serious constraint than the necessity to have a new pair of glasses. Subjects probably made such a judgment, both on the basis of their general world knowledge and on the basis of information implied in the story (no information was stated about Mr. Bianchi's myopia, so that subjects could infer that he did not really need the new pair of glasses). Thus, differential availability of changes seems to depend on the relative degree of control that subjects attribute to events which all correspond to some intentional decision of the protagonist of the story.

Our argument is based on the intuitive distinction between controllable and uncontrollable events. However, in the paradigm of undoing a scenario, subjects do not control, obviously, the course of events, not even those representing the protagonist's actions. Therefore, it might be argued that it is not well-founded to conclude that people consider more mutable the events that are under their *direct* control. This point can be taken as a legitimate criticism of using this paradigm, and as a suggestion for investigating mental simulation processes in a more ecologically valid way. (It might be fruitful, for example, to conduct experiments in which people are actually engaged in some activities.) However, in recent psycholinguistic research on reading, there are some persuasive, although indirect, pieces of evidence for our conclusions. Work on the comprehension of narratives showed that 'readers take the character's perspective; they follow the character's thoughts, activating mental images of the same things that the character is thinking about' (Bower and Morrow 1990: 47). And, for example, Morrow et al. (1989) found that, in drawing spatial inferences, subjects focus more on the character's mental location than on his/her physical location if the former is more relevant to the character's current plan. Although these results were obtained by using a paradigm different from the undoing of scenarios, they clearly confirm our interpretation. Both in constructing a mental model of the situations described in a narrative (cf. Johnson-Laird 1983), and in simulating a new outcome of it, readers seem to focus on information relevant to the protagonist, drawing inferences from his/her specific point of view.

General discussion

Our general hypothesis was that in the process of mentally undoing the dramatic outcome of a scenario by mutating its component events, a relevant factor is the perceived degree of control that the protagonist can exercise on these events. In the present experiments, we manipulated the controllability of events by presenting scenarios containing events that in some cases were under the control of the protagonist, and, in other cases, were out of his or her control. Taken together, the present results provide strong support for our hypothesis by showing that the former events are more mentally mutable than the latter ones.

In experiment 1, subjects had to undo mentally the dramatic outcomes of scenarios in which the protagonist made some intentional decisions and had some minor misfortunes that did not depend on his free will. The results showed that people prefer to alter the events corresponding to the protagonist's decision, regardless of the events' position in the causal chain. Although the causal order effect predicted by Wells et al. (1987) was replicated, the controllable events were more altered than uncontrollable ones, even when presented in the last position. The present finding does not necessarily imply that the position in a sequence of events is an irrelevant factor in determining their relative mutability. (For example, it is likely that subjects would more readily alter the first rather than the subsequent actions of a causally ordered sequence of *equally controllable* actions). Moreover, our argument (that controllable events are more mutable than uncontrollable ones) does not conflict with Wells et al.'s argument (that effects are somewhat immutable because they are constrained by prior causes). According to Wells et al. (1987: 426), 'prior causes are [...] perceived as having greater freedom of occurrence and nonoccurrence' than effects. Now, from the protagonist's point of view (which is the one taken by the subject), his/her own actions have a greater freedom of occurrence and nonoccurrence than external events (which, by definition, occur independently of his/her free will). Therefore, a smaller number of constraints makes causes more mutable than effects, as well as controllable actions more mutable than external events.

In experiment 2 we found that the normal action of the main character is more readily mutated than more exceptional external events. Even though these events are not exceptional per se, it should be emphasized that in the context of the story they were certainly more exceptional than the normal action of drinking a beer in the bar. Contrary to our predictions, based on the assumed asymmetry between norms and exceptions (Kahneman and Tversky 1982a), the exception status attributed to the protagonist's action did not increase its mutability much. Of course, the failure to find a significant difference between the normal and the exceptional action cannot be taken as evidence that there is no asymmetry between norms and exceptions. In experiment 2, the ceiling effect produced by the high mutability of the normal action is probably the result of the high availability of some, although unspecified, counterfactual alternatives. In other words, the controllable action, being the product of an individual's decision, evokes, by definition, the possible action(s) that the individual did not undertake. Future work may clarify the relationship between the degree of normality and the degree of controllability of events in determining their mutability. In the meanwhile, the present finding corroborates our assumption about the basic asymmetry between controllable and uncontrollable events.

Apart from an experiment by Legrenzi et al. (1984), the present one is the first mental simulation experiment in which an exceptional outcome was altered by changing a normal event. Legrenzi et al. (1984)

presented their subjects with a story with a tragic outcome (a collision between a train and a car at a level-crossing). The protagonist of this story (the level-crossing keeper) was stopped by an exceptional event (a bridge falling down) on his way to the level-crossing. Despite the exceptionality of the main event of this story, most of the subjects (60%) preferred to change the usual time for the keeper to leave his house, in the direction of exceptionality (e.g., 'If only he had left home earlier that night ...'). Only a minority of subjects (6%) mentally eliminated the bridge falling down. The original aim of this experiment was to show that, contrary to Kahneman and Tversky's (1982a) predictions, people make mutations along a continuous variable (time), when requested to undo an outcome. The results by Legrenzi et al. can be taken as a piece of evidence in favour of our general hypothesis: People changed the only controllable action of the story. In particular, these results are consistent with those of experiment 2, by showing how normal controllable events can be readily mutated toward exceptionality to undo an exceptional outcome.

Recently, Gavanski and Wells (1989) found that subjects primarily altered exceptional events in the direction of normality to undo exceptional outcomes, but changed normal events in the direction of exceptionality to undo normal outcomes. They explain these results in terms of the correspondence heuristic, which refers to the belief that exceptional outcomes have exceptional causes and that normal outcomes have normal causes (this belief is clearly included in the general belief that causes and effects are similar, cf. Einhorn and Hogarth 1986). Although it is likely that this heuristic might often guide the mental simulation, it does not seem to be its primary determinant. Our results indicated that the correspondence effect is not present in situations in which there are relevant differences in the degree of event controllability. In these situations, even a *normal* controllable event (leaving home, drinking a beer) is more altered than more exceptional but uncontrollable events (a bridge falling down, a tree trunk on the road), to undo an *exceptional* outcome. Thus, the way in which the principle of correspondence determines the construction of counterfactual scenarios depends critically on the nature of the events stated in the initial scenarios.

In experiment 3, we studied the relative role of focus and controllability of events on mental simulation. Despite a common reference to the focus of the story, three actions that depended to a different degree on the protagonist's free will turned out to be differently mutable. This finding is consistent with the results of a study by Kahneman and Miller (1986: 143). They presented their subjects with the following question: 'Tom and Jim both were eliminated from a tennis tournament, both on a tie-breaker. Tom lost when his opponent served an ace. Jim lost on his own unforced error. Who will spend more time thinking about the match that night?' The results were clear-cut: most of the subjects (85%) answered that Jim was more upset than Tom. According to Kahneman and Miller, these results are due to the differential availability of counterfactual alternatives for events that can be changed by introducing an improvement or a deterioration. In this case, the same undesirable outcome (to be eliminated from the tournament) is judged to be more upsetting when it involves an imagined improvement of performance (to avoid the error). It is likely that the imagined improvement and degradation of one's own performance is a factor that governs mental simulation. It should be noted, however, that the results obtained with the tennis story could easily be explained in terms of different degree of control. Since in this task the subjects are implicitly taking the protagonist's point of view they are considering, for Jim, the possibility of changing his own performance, that is, they are making mutations for a controllable action. By contrast, for Tom, they are considering the possibility that someone else (Tom's opponent) changes (i.e., deteriorate) his/her performance. This means that the subjects are making mutations for an action which is uncontrollable from Tom's point of view. In sum, these results suggest that the improvement/degradation effect can be considered as a special case of the more general controllability effect. We discuss these results here because they clearly support our argument about focus. As in our case, despite a common reference to the focus (both Tom and Jim were focal actors in the tennis story), events that were represented as more controllable elicited more counterfactual alternatives.

Ergonomic implications

The present results show that the subjective controllability of an event makes it highly mutable. Given the relationship between mutability of an event and perceived causality (Wells and Gavanski 1989), the human action is likely to become the 'figure' on the 'ground' of the other events when it has to be established how a given output was produced. The subjective controllability can be relevant for the assessment of blame and causal roles in accidents and the consequent feeling of regret and guilt of operators. Operators involved in a malfunction could experience a feeling of guilt and/or could be blamed for their performance just as a function of the availability of alternatives to their actions. This quite apart from the real possibility to modify the human performance in that context. Vice versa, operators might feel themselves 'innocent' if they were convinced that there nothing could be done under their control to avoid the breakdown. Most interesting, in man–machine interaction it frequently happens that humans compensate under-functionalities of the system with their adaptive behavior. It becomes thus possible to keep constant the qualitative outcome of the interaction (Perrow 1984). Very often this compensation is not occasional, but is a long-lasting process that modifies the original balance between man and machine. Both in bad system design and in functional machine degradation, humans are prone to compensate under-functionalities with their behavior. Everyone who becomes gradually used to drive an old car is involved in this type of process. The driver not only develops special skills but gradually takes in charge mechanisms once belonging only to the car. Consider accidents occurring to this type of driver. Paradoxically, here, the human performance involved in the accident is even more prone to be considered as the event that could be avoided. This attribution of responsibility is likely to be done according to the same counterfactual thinking analyzed in our experiments. Indeed the driver could control his/her skills (action to be modified) but not the car's malfunctioning of the car (out of his/her control). In the same way, remedies to error-prone scenarios could be subject to this bias. That is, the causal role attributed to operations under human control will be the focus for modifying the man–machine system. This line of reasoning clearly supports Norman's (1990) criticisms to the 'blame and train' philosophy widespread among designers. Man rather than machine is responsible for error, and human action is the 'easiest' factor to modify.

A more general question is whether these results and the counterfactual processes can be applied to the analysis of every type of error and deleterious outcome. We can consider the triple categorization of errors discussed by Rich (1983). She distinguishes between ignorable, recoverable and irrecoverable errors. An ignorable error has no consequences on the state on which the operator is acting upon. For example if you switch on a computer screen with your pointing device where there is no button, there are no modifications at all in the state of the system. You can simply replace that action with another action. The only consequence might be a waste of time (here it is assumed that such time-consuming activities are irrelevant to the main activity). A recoverable error has consequences on the state of the system that does not fit with the aim of the activity. Here the operator should try to undo the undesirable state the system has assumed. This could be done either by restoring the previous state, or by getting some intermediate state from which it is feasible to attain the desired state. For example, if you decide to clean your floppy of an undesired file and you select a wrong file and delete it, you need to undo the operations. In order to do that you recover the file from the basket. Finally, an irrecoverable erroneous solution simply cannot be undone. For example, adopting the previous example, you could have turned off the computer before you realized the erroneous selection. You cannot go back to the previous state of the system and undo the erroneous operations. In most cases the study of counterfactual thinking concerns irrecoverable negative outcomes. It would be interesting to study the mechanisms of counterfactual thinking in the two other cases. In the ergonomic literature some studies have investigated the 'near-misses' situations in which an accident has been avoided at the last moment (e.g., Malaterre and Muhlrad 1976). In the literature on counterfactual thinking there is a study in which a story with an irrevocable dramatic outcome was compared to a story in which the dramatic outcome was miraculously avoided. Legrenzi et al. (1984) modified the. outcome of Kahneman and Tversky's (1982a) story about Mr. Jones. In this modified version, Mr. Jones' accident was avoided miraculously at the last moment. In this case, subjects preferred to modify the timing of the accident. By contrast, in the version with the negative outcome, they preferred to modify the exceptional action (Mr. Jones' change of his usual route). These results suggest that people apply different strategies when they have to simulate alternatives to irrecoverable accidents and to near-accidents. If this hypothesis is correct, it would be interesting to study the subjective and phenomenological categorization of errors and their relationship with counterfactual thinking. For example, a recoverable error could be judged irrecoverable as a consequence of the incapacity to imagine the *right* alternative to the action that caused the error. It is likeky to be the nature of counterfactual alternatives that makes the accident *subjectively* ignorable, recoverable or irrecoverable.

References

Bower, G.H. and D.G. Morrow 1990. Mental models in narrative comprehension. Science 247, 44–48.

Dennett, D.C., 1984. Elbow room. The varieties of free will worth wanting. Cambridge, MA: MIT Press.

Dunning, D. and M. Parpal 1989. Mental addition and subtraction in counterfactual reasoning: On assessing the impact of personal action and the life events. Journal of Personality and Social Psychology 56, 5–15.

Dupuy, J.P., 1990. Temps du projet et temps de l'histoire [Time of the project and time of the history]. Technical report n. 9014A. CREA, Ecole Polytechnique, Paris, France.

Einhorn, H.J. and R.M. Hogarth 1986. Judging probable cause. Psychological Bulletin 99, 3–19.

Fabre, J.M., in press. La relativisation des jugements [The relativity of judgement]. Doctoral dissertation, University of Provence, France.

Gavanski, I. and G.L. Wells 1989. Counterfactual processing of normal and exceptional events. Journal of Experimental Social Psychology 25, 314–325.

Johnson, J.T., 1986. The knowledge of what might have been: Affective and attributional consequences of near outcomes. Personality and Social Psychology Bulletin 12, 51–62.

Johnson-Laird, P.N., 1983. Mental models. Cambridge: Cambridge University Press.

Kahneman, D. and D.T. Miller 1986. Norm theory: Comparing reality to its alternatives. Psychological Review 93, 136–153.

Kahneman, D. and A. Tversky 1982a. 'The simulation heuristic'. In: D. Kahneman, P. Slovic A. Tversky (eds.), Judgment under uncertainty: Heuristic and biases. New York: Cambridge University Press. pp. 201–208.

Kahneman, D. and A. Tversky 1982b. The psychology of preferences. Scientific American 246, 160–173.

Kahneman, D. and C.A. Varey 1990. Propensities and counterfactuals: The loser that almost won. Journal of Experimental Social Psychology 59, 1101–1110.

Kelly, H.H., 1967. 'Attribution in social psychology'. In: D. Levine (ed.), Nebraska Symposium on Motivation, Vol. 15. Lincoln, NE: University of Nebraska Press.

Landman, J., 1987. Regret and elation following action. Personality and Social Psychology Bulletin 13, 524–536.

Legrenzi, P., R. Rumiati and M. Sonino 1984. L'euristica della simulazione [The simulation heuristic]. Report no. 109, Istituto di Psicologia, University of Padua, Italy.

Leplat, J., 1987. 'Accidents and incidents productions: Method of analyis'. In: J. Rasmussen, K. Duncan and J. Leplat (eds.), New technology and human error. Chichester: Wiley.

Lipe, M.G., 1991. Counterfactual reasoning as a framework for attribution theories. Psychological Bulletin 109, 456–471.

Malaterre, G. and N. Muhlrad 1976. Intérêt et limite du concept du conflit de traffic et quasi-accident dans les études de sécurité. [Validity and invalidity of the concept of traffic conflict and near-accident in the security research]. Technical Report, ONSER, Monthléry.

Miller, D.T. and S. Gunasegaram 1990. Temporal order and the perceived mutability of events: Implications for blame assignement. Journal of Personality and Social Psychology 59, 1111–1118.

Miller, D.T. and C. McFarland 1986. Counterfactual thinking and victim compensation: A test of norm theory. Personality and Social Psychology Bulletin 12, 513–519.

Miller, D.T., W. Turnbull and C. McFarland 1990. 'Counterfactual thinking and social perception: Thinking about what might have been'. In P. Zanna (ed.), Advances in experimental social psychology, Vol. 23. pp. 305–331. Orlando, FL: Academic Press. pp. 305–331.

Morrow, D.G., G.H. Bower and S.L. Greespan 1989. Updating situation models during narrative comprehension. Journal of Memory and Language 28 292–312.

Norman, D.A., 1990. Cognitive science in the cockpit. Paper presented at the 'Aerospace Human Factors Symposium', NASA-Ames Research Center, Moffet Field, CA, April 11, 1990.

Perrow, C., 1984. Normal accidents. New York: Basic Books.

Read, D., 1985. Determinants of relative mutability. Unpublished research, University of British Columbia, Vancouver, Canada.

Rich, E., 1983. Artificial intelligence. New York: McGraw-Hill.

Sappington, A.A., 1990. Recent psychological approaches to the free will versus determinism issue. Psychological Bulletin 108, 19–29.

Turnbull, W., 1981. Naive conception of free will and the deterministic paradox. Canadian Journal of Behavioural Science 13, 1–13.

Wagenaar, W.A., P.T.W. Hudson and J.T. Reason 1990. Cognitive failures and accidents. Applied Cognitive Psychology 4, 273–294.

Wells, G.L. and I. Gavanski 1989. Mental simulation of causality. Journal of Personality and Social Psychology 56, 161–169.

Wells, G.L., B.R. Taylor and J.W. Turtle 1987. The undoing of scenarios. Journal of Personality and Social Psychology 53, 421–430.

Computer programming
and program debugging

Acta Psychologica 78 (1991) 137–150
North-Holland

Training of Pascal novices' error handling ability *

Carl Martin Allwood and Carl-Gustav Björhag

University of Göteborg, Göteborg, Sweden

The present study reports on the effects of providing novices in Pascal one hour of training in the debugging of their own programs. The training presented subjects with results from previous research on novices' debugging as well as fairly detailed instructions on how to debug a program. Furthermore, the instructions included explanations of common error messages from the computer. The debugging performed by the eleven subjects who received training was compared with the debugging performed by ten control subjects who programmed on their own during the training hour. The results show that the subjects who received training debugged a significantly larger proportion of their errors compared with subjects who received no training. The results suggest that especially the semantic programming errors were easier to debug for the trained group. No differences in detailed debugging behaviour were found between the groups.

Introduction

Debugging of computer programs is important because it can help prevent accidents and other unfortunate events [1]. Furthermore, previous research shows that debugging occupies a substantial part of programmers' time. For novice programmers, two studies (Allwood and Björhag 1990; Miller 1974) indicate that debugging can take up at least half of the time spent programming. From a theoretical perspective, debugging is interesting as an example of evaluative processes. During evaluative processes the individual has an opportunity to stand back, reflect and possibly change his or her understanding or actions from a more deliberated perspective. Considering the practical importance of

---

* Requests for reprints should be send to C.M. Allwood, Dept. of Psychology, University of Göteborg, Göteborg, Sweden.

[1] The term 'debugging' is here, as in the rest of the paper, used in the general sense of searching and correcting errors in programs, i.e., it is not used in the more technical sense of searching for errors by means of a symbolic debugger.

debugging it is surprising that most textbooks on programming contain only a few pages on how to debug computer programs.

The importance of debugging clearly indicates the desirability of developing methods for training novices to debug computer programs. Research concerned with the development of methods or aids and/or the comparison of various methods or aids for debugging appears to be somewhat more common (for example, Myers 1978; Gilmore and Smith 1984) than research dealing with training of debugging skills. In fact, we have been able to locate only two studies (Carver and Risinger 1987; Katz and Anderson 1988: exp. 4) which evaluate training in debugging. The study by Carver and Risinger demonstrated the effectiveness of a training regime in debugging developed for sixth graders' programming in LOGO. Half an hour of instruction in debugging resulted in improved debugging for the experimental group as compared to a control group.

The instructions in the training given by Carver and Risinger emphasized the importance of a step-by-step approach when debugging, going from more general questions to more specific ones. When debugging the student was first to ask him/herself, 'What is the problem?' and then, 'What type of bug could cause the problem?' Next, the student was instructed to decide if the program to be debugged had subprograms and, if so, if the bug might be located in a specific subprogram, in a REPEAT- or an IF-construction, or after a specific command. If none of these questions helped, the student was asked to read every command and decide whether or not it was correct. Finally, after having corrected the bug, the student was asked to re-test the program.

The students' debugging skills were tested 'shortly after the debugging instruction' (Carver and Risinger 1987: 159). In the test-phase, the students were asked to debug two programs written by somebody else but that they had previously tried to code themselves. Each of the two programs contained about 20 lines with four clusters of five lines having nearly identical structures. The students were instructed 'to fix all the bugs' (p. 158). Each program contained eight planted bugs and the students were allowed 'one class period' to debug each program.

Katz and Anderson (1988) used planted bugs and compared the training of three different ways to locate a buggy line in programs of about ten lines. Subjects' bug-location strategy was manipulated through the use of different debugging interfaces which forced the subject to use a specific bug-location strategy. The first bug-location strategy was to have subjects go through each line of code from the start in the order the program would be executed by LISP. The second bug-location strategy was to have subjects locate a program line to look at starting from the buggy output of the program. The third bug-location interface involved asking the subjects directly which area in the program they would like to look at. For all three conditions, the subjects were asked to judge any line they looked at with respect to whether or not it was correct, and to supply a correct version of the line if the line was judged to be erroneous.

The results showed that forcing the subjects to use a specific bug-location strategy during the debugging of six short programs resulted in the subjects' tending to use the same bug location strategy when debugging six further programs with the exception that the subjects in the first condition read the programs in prose order rather than in execution order. However, there was no clear relation between bug-location strategy and correctness of judgements as to whether a certain line of code was correct or not, nor with respect to replacing a buggy line with a correct new line or not.

In common with the study by Carver and Risinger (1987) and by Katz and Anderson (1988: exp. 4), most studies on debugging have dealt with the debugging of bugs planted by the researcher. However, debugging of computer programs may also take place in a context where the program to be debugged is written by the same person who is doing the debugging. Both of these situations are presumably common in programming contexts. As suggested elsewhere, there may be important differences between the cognitive processes involved in each of these two occasions (Allwood 1986; Allwood and Björhag 1990; Katz and Anderson 1988; Waddington and Henry 1990).

For example, Katz and Anderson (1988: exp. 3) compared the debugging of subjects' own LISP-programs and the debugging of LISP-programs created by others. In order to create comparability between the two conditions (own and other produced programs), different constraints were imposed on the subjects' debugging. In spite of these constraints, which, as the authors note, work against finding differences between the two conditions, differences were still found between the conditions. For example, subjects debugging their own programs found more of the bugs and required less time to do so than when they debugged programs written by others. Furthermore, the two conditions differed in the debugging strategies employed by the subjects. Debugging of self-produced programs involved less time used in trying to understand the program. It was also found to involve more use of backward-reasoning, i.e. a bug location strategy starting from the buggy output of the program, as opposed to forward-reasoning, i.e. reading and evaluating the program line by line from the start.

A comparison between two studies by Gugerty and Olson (1986) and Allwood and Björhag (1990) also indicates that comprehension processes play a smaller role when debugging self-produced programs as compared with other-produced programs. Gugerty and Olson (1986) used planted bugs and compared the debugging of experts and novices. Their results indicated that the experts' advantage in debugging was due to their better skills in program comprehension. Allwood and Björhag (1990) studied novices' debugging of their own programs and in this context found evidence of limitations to the importance of program comprehension processes in the debugging phase.

Furthermore, studies of debugging of planted bugs usually do not include the stage where the programmer goes from not suspecting a bug to supposing that the program is buggy (Allwood 1986). It is also the case that studies of debugging using planted bugs generally use fairly simple bugs and, moreover, the remaining program, if written by the researcher, tends to have a good structure. In contrast, the bugs in studies using self-produced programs can be complex and programs of any substantial length written by novices often have a poor structure. Finally, novices debugging their own program may be more susceptible to effects of einstellung and functional fixedness since they are working with code produced by themselves. An analogy can be made to the often reported contrast between the ease of critically reading a text by another person and the difficulty of critically reading your own text.

Waddington and Henry (1990) report results which indicate that experts, when debugging programs written by others, as part of their debugging strategy attempt to reconstruct the intentions of the program's author.

Debugging is clearly a complex cognitive activity, the details of which are dificult to study. In order to structure the situation, nearly all researchers have attempted to simplify the situation in various ways. Such simplification often involves putting restrictions on what the subject may do when attempting to debug a program. For example, in many studies subjects were not allowed to interact with the computer when attempting to debug the program (for example, Adelson 1984;

Vessey 1985; Waddington and Henry 1990; Wiedenbeck 1985). It appears likely that the same person will use different strategies for debugging with access and without access to a computer.

Other studies have used very small programs (sometimes less than ten lines); have only studied a limited part of the debugging process; have introduced rules for the debugging strategies subjects may use; and have automatically eliminated corrections which did not eliminate the bug, etc (e.g. Katz and Anderson 1988). The effect of the introduction of such limitations on subjects' debugging is not clear. Moreover, in most studies, subjects' debugging processes have not been followed step-by-step but instead other types of data have been collected, for example subjects have been asked to give retrospective reports on their debugging.

The present study is an attempt to study the effect of debugging training in a somewhat more realistic situation than those used in most earlier studies of debugging. We tested the effect of our debugging training by letting the subjects attempt to create a bug-free program, a correct version of which amounted to approximately 40 lines of code.

The content of the training was to a large extent derived from the weaknesses in subjects' debugging processes evidenced in the study of Allwood and Björhag (1990) and other studies reviewed in Allwood and Björhag (1990), such as Davis (1983) and Miller (1974). For example, Allwood and Björhag (1990) found that 32% of the bugs indicated by error messages from the compiler were not eliminated by the subjects. Some reasons found for this were that the subjects sometimes ignored the error messages, that the error messages were occasionally misleading and that the subjects did not always correct the bugs in the order in which they appeared in the program. The last mentioned fact led the subjects to attempt to correct 'errors' which stemmed from the fact that the compiler had parsed the program erroneously because of an earlier bug in the program.

Miller (1974) used questionnaire data to investigate novice programmers' debugging strategies. The results showed that the debugging strategies employed by the novices were passive in contrast to the more active debugging strategies employed by professional programmers. Davis (1983) found that novices debugging their own programs had difficulty in understanding the implications of the error messages given by the system.

In brief, our training attempted to activate the subjects, to structure their behaviour and to provide them with information about the content of the messages given by the computer.

Method

Subjects

Twenty-two undergraduate students from the Department of Industrial Economy at the Chalmers University of Technology, Göteborg, Sweden participated in the study. All were novices in programming and had just completed the first seven weeks of an introductory course in Pascal.

No clear differences were found between the subjects with respect to previous experience in programming or with respect to their rating on a five-step scale of how much fun they thought it was to program. Three subjects in each group did not have any prior programming experience, most of the remaining subjects had only taken courses in Basic in school at the pre-university level. Only four subjects had a computer at home (three in the control group); for the most part the computer was used for games.

All subjects were volunteers and were paid for their participation. Eleven subjects were randomly assigned to an experimental group and 11 to a control group. One of the subjects in the control group was later deleted from the study due to technical problems with the video recording of the subject's screen.

Materials

Programs
The version of Pascal used in the study was a version of Standard Pascal, running on a NORD500 computer under the operative system Sintran. The text-editor used when entering the code into the computer was an editor called Ped.

Training instructions in debugging
The training instructions were five pages long. The first section of the text presented results from previous research (our own and others) on deficiencies in novices' debugging of their own code. Novices' debugging was described as generally being too passive. Common deficiencies mentioned were that novices often do not (1) sufficiently attempt to follow the flow of information in the coded program, (2) attempt to understand the content of the error messages given by the computer, (3) bring up the program on the screen in order to find the place in the program referred to in the error message, and (4) sufficiently attempt to understand the effects of the corrections that they plan to carry out in the program before the corrections are made.

The subjects were also warned that the compiler might sometimes parse the program in an incorrect way and because of this give erroneous information about the bugs. However, it was stated that the error messages commonly give very useful information.

The subjects were further informed that when novices' debugging has been compared to that of experts it has been found that novices more often engage in debugging activities which are less intellectually demanding. Furthermore, research results were described which show that novices often debug their programs in an unsystematic way, for example by attempting to debug errors occurring later in the program before correcting errors occurring earlier in the program. This might result in the correction of 'errors' (code incorrectly parsed by the compiler) which would have disappeared by themselves had the earlier occurring bugs been corrected. The unsystematic and mechanical use of test values by novices when testing the program was also commented on.

In the next section of the instructions the novices were asked to debug their programs according to the following instructions (translated from Swedish):

'(1) Start to correct the bugs in the order in which they are marked in the program (i.e. not in the order they are given in the compiler's "error list"). Try not to deal with too many bugs at the same time; preferably deal with only one bug at a time (you might get rid of a few sequel bugs at the same time). When you start to correct a bug, first read the error message dealing with the bug and explain the content of the error message.

(2) When you have understood the meaning of the error message for a certain bug but you do not immediately realize what the error is, bring up the program on the screen. Next try to localize the area in the program which the compiler might have indicated. Then look for syntax errors if the text of the error message gives you reason to suspect this. Examples of syntax errors are: omission of a semicolon (;), one semicolon too many, and the use of loop-constructions which are not grammatically correct.

If the error message indicates other types of bugs try to trace these by carefully following the flow of information from the beginning of the program. If this seems completely unnecessary, start to follow the flow of information a little above the place in the program suggested by the compiler as the place of the bug. For each program line that you read in the program state aloud what is carried out in that line. Then consider if it fits in with what the program did earlier and if it fits with what you planned that the program should do.

(3) Before you correct a bug, carefully make it clear for yourself what the nature of the bug is. Before you carry out a correction you should make yourself aware of what consequences the change will have on the rest of your program. Also make sure your change does not violate any of the assumptions which the program is dependent on.

(4) If you have corrected a bug which, in turn, might have given rise to other bugs you should immediately investigate the effect of your correction by compiling or running the program in order to find out if your change has had the intended effect.'

In the next section the subject was given information about the local compiler. This information included information about bugs which diminished the compiler's chance of interpreting the program. For example, it was pointed out that a semicolon directly preceding an 'ELSE' usually leads to the rest of the program being completely misinterpreted by the compiler. Two more such examples were given together with typical error messages indicating the presence of such bugs.

In the final section of the instructions the meanings of 31 common error messages concerning compile-time errors were explained as were the meanings of four common messages concerning run-time errors.

Problems for the control group

Four small programming problems were prepared for the subjects in the control group. The simplest of these problems was to create a program adding five numbers given to the program by the user from the keyboard. The most complicated of the four problems was to construct a program which could read in values and compute the mean and the standard deviation of these values. The formulae for the mean and the standard deviation were given in the instructions in statistical notation.

Training programs

Solutions to the above mentioned four problems were planted with approximately five bugs each. The bugs were of syntactic, semantic as well as logical types (these error types are defined below). These buggy programs together with the corresponding problem instruction texts given to the control group were used as part of the training material in the experiment group.

Test problem

This program was to calculate the sum of a simple mathematical series (the harmonic series) for any given value of *n*. The task included all stages of programming, including debugging and testing the program. This was a slightly simplified version of the problem used by Allwood and Björhag (1990). The instructions to the problem are given in the Appendix and a possible correct solution to the slightly more difficult problem is given in Allwood and Björhag (1990). A correct solution to the present problem is about 38 lines long. All problem instructions were given in writing.

Procedure

All subjects were first given an overview of the session and were then asked to fill in a brief questionnaire about their programming experience. In the subsequent one-hour training session the subjects in the experimental group were asked to read the debugging training instructions, described above.

After having read through the instructions carefully and having been given the opportunity to ask questions, subjects in the experimental condition were presented with the four training programs, one at a time. As noted above, each training program had planted bugs and the subjects were asked to retrieve the programs from a prepared file to the screen, compile and debug the program so that it would function according to the description given in the problem instruction. The subjects were asked to debug the programs according to the debugging instructions just read. The experimenter intervened during the training session to remind the subject to debug according to the training instructions, should he or she forget to do so. The subjects in the experimental condition had free access to the debugging instructions during both the training phase and the test phase.

In the control's training session subjects were asked to create correct programs for each of the four training problems. In contrast to the task given to the experimental subjects, this task included writing the actual code from the beginning.

In the test phase, which lasted for two hours, all subjects were asked to create a correct version of the test problem described above. All subjects were given the same instructions except that the subjects in the experimental condition were asked to debug according to the training instructions. The instructions in both conditions emphasized that the subject should attempt to solve the problems with as little help from the experimenter as possible. However, subjects were also told that the experimenter would ask them if they wanted help if they got stuck and sat passively for a few minutes. A *t*-test analysis showed that there was no statistically significant difference between the conditions with respect to the number of times they were aided by the experimenter ($p > 0.42$).

All subjects were allowed free use of their text book in Pascal. The subjects in both conditions were asked to verbalize their thoughts aloud ('think-aloud') during the whole test phase. The subjects were also asked to report what commands they entered on the computer. The computer screen was videotaped and the subjects' think-aloud comments were tape-recorded during the test phase. Each subject was seen individually.

Results

The raw data used in the following analysis consisted of the video recordings made of the subjects' screen as they attempted to solve the programming task in the test phase of the experiment. The raw data also included the tape recordings of subjects' think-aloud reports and notes taken by the experimenter during the test phase. The data analyses presented here only relate to events after the subjects had attempted to compile the program for the first time. This means that any bugs debugged before the first compilation of the program are not included in the analyses presented below, all other bugs are included. However, previous research indicates that most of novices' debugging activities usually take place after the subject has attempted to compile the program for the first time (Allwood and Björhag 1990; Gray and Anderson 1987).

Bugs made

Four types of programming errors were analyzed; spelling errors, syntax, semantic and logical errors. Syntax errors are defined as program code which does not follow the grammatical form necessary for the running version of Pascal. Syntax errors also include instances where the subject omits code which is obligatory in the program, such as declarations of variables used in the program. Semantic errors are syntactically correct expressions which are impossible in the context they occur in. Finally, Logical errors are errors where the code is syntactically correct but does something different

Table 1
Mean number of errors made in the two conditions, for four types of errors and for all errors.

Error	Condition		df	p-value
	Experimental	Control		
Syntax	5.24	5.50	19	0.90
Semantic	1.91	1.30	19	0.33
Logical	4.27	3.90	19	0.69
Spelling	0.45	0.40	19	0.88
All errors	11.91	11.10	19	0.74

than is stated in the problem instruction. In the few cases where errors were nested, i.e., a syntax error occurring in a logical error, these were analyzed as separate errors.

Table 1 shows that the two conditions do not differ with respect to the total number of bugs made nor with respect to the number of bugs made when the four types of bugs analyzed are considered separately.

Bugs eliminated

Table 2 shows the mean proportion of corrected errors with respect to all syntax, semantic and logical errors made in each of the two conditions and to all errors made in each condition. As can be seen in the table, there was a significant difference between the conditions with respect to the proportion of errors corrected by a subject to all errors made by the same subject. The table also shows that the proportion corrected Semantic errors to all Semantic errors made approached significance. There was no statistically significant difference between the groups with respect to the number of errors introduced after the first compilation of the program.

Debugging behaviour

The next analysis deals with possible differences between the two conditions with respect to the subjects' debugging behaviour. For this purpose the subjects' behaviour

Table 2
Mean proportion of corrected errors with respect to all syntax, semantic and logical errors made and to all errors made for the two conditions.

Error	Condition		df	p-value
	Experimental	Control		
Syntax	0.99	0.95	19	0.36
Semantic	0.96	0.67	19	0.09
Logical	0.87	0.75	19	0.24
All errors	0.94	0.83	19	0.02

in those parts of the testing-phase where the subjects evaluated their program in a negative way was coded by one coder with a slightly modified version of the coding scheme used by Allwood and Björhag (1990). After each coding category, below, we give (in parentheses) the average number of occurrences of the category for a subject, first for the experimental condition and then for the control condition.

(1) Interprets Meaning of Error Message. The subject interprets the meaning of an error message; for example, the subject rephrases the content of the error message (6.7; 5.8).
(2) Follows Flow of Information in the Program. The subject mentally 'simulates' or 'interprets' some part of the program by trying to imagine the flow of information in that part (1.5; 1.0).
(3) Error Hypothesis Localized to Construction. The subject hypothesizes about what might be wrong with the program. The hypothesis is limited to the mention of a specific construction, i.e. a FOR-loop or a part of the program carrying out a specific function such as controlling for overflow (3.0; 2.9).
(4) Exact Error Hypothesis. The subject's hypothesis about the error specifies the exact code which is thought to be buggy (6.5; 5.8).
(5) Compiles. The subject compiles the program (10.9; 13.6).
(6) Tests. The subject enters a value into the program and observes the program's response (29.7; 25.5).
(7) Enters program on screen. The subject enters the program on the screen (8.6; 12.4).
(8) General Dissatisfaction. The subject expresses general feelings of helplessness and dissatisfaction with respect to the program (3.6; 5.8).
(9) Enters evaluative phase. The subject stops producing code for a new section of the program and starts to evaluate previously produced code (2.8; 2.1).

Statistical analyses (using *t*-tests) showed no statistically significant differences between the two conditions with respect to any of the categories investigated.

Discussion

The present study involves an attempt to improve subjects' debugging of a program created by themselves by giving them training in debugging. The main result of the study was that the experimental group after one hour of training in debugging eliminated a significantly larger proportion of their bugs compared to a control group which had received no such training. There was no significant difference between the conditions with respect to the number of bugs made. This suggests that the groups were equivalent with respect to their knowledge of Pascal.

The result that debugging of own produced programs can be improved after training together with the fact that novices spend much time debugging their programs suggests the desirability of introducing debugging training into the curriculum in programming courses.

An important question in connection with training in debugging concerns which components of the training were effective and which were not. Unfortunately we were not able to locate any systematic differences between the two conditions with respect to debugging behaviour. The lack of differences in this respect between the two conditions in our study possibly suggests that our debugging training did not serve to increase the general level of activity of subjects in the experimental condition.

For some of the categories investigated, the fact that subjects in both groups were asked to verbalize all thoughts aloud may have contributed to the lack of differences found between the two groups in specific debugging behaviour. For example, the think-aloud instructions may have encouraged the subjects in the control condition to become more explicit about the meaning of error messages, just as was encouraged in the debugging instructions. Furthermore, our analyses only dealt with the number of times a specific category occurred; it may be that there are differences between the conditions as to the order in which and quality of how the debugging behaviours were carried out.

Both in the present study and in that of Carver and Risinger (1987) the debugging training given encouraged subjects to be systematic in their debugging behaviour. It may be that the beneficial effects of the training to a large extent were located on this general level.

In order to develop more efficient debugging training programs it is important to find out exactly what training components will produce effective debugging behaviour. In this connection we assume that it will not always prove efficient to attempt to make novices copy the debugging behaviour of experts. For example, Nanja and Cook (1987) found that the experts in their study often corrected many bugs at the same time before verifying the effects of their corrections. In contrast, the novices in the same study tended to correct one bug at a time. When the novices in the study by Allwood and Björhag (1990) attempted to correct many bugs at the same time they occasionally got into difficulties. It seems likely that the greater processing capacity available to experts in connection with programming allows them to successfully use debugging strategies which are too resource demanding for novices.

Further research is needed to make clear what training components result in efficient debugging behaviour.

Appendix

Programming task given to subjects.

Harmonic series

The task is to write a PASCAL program which calculates the sum H of the so-called harmonic series.

$$H = 1 + 1/2 + 1/3 + \ldots + 1/n, \quad \text{where } n > = 1.$$

The obvious way to calculate this sum is to add the n terms involved. However, for large values of n this way of doing the calculations will be very time-consuming, even for a fast computer. For large n-values, the sum 'H-approx.' can be used as an approximation. The formula for this is:

$$H\text{-approx.} = \ln n + Y + 1/(2n) - 1/(12n^2) + 1/(120n^4).$$

where the constant $Y = 0.5772\,1566\,4901\,5329$ (rounded to 16 decimals).

When $n > 200$ the difference between the two formulas is less than 10^{-16} (i.e. 16 decimals, which is the approximate limit for the computer's precision using 'longreal'). Thus, for $n > 200$ the approximation formula should be used.

The numerically largest number, 'MAXREAL', which can be represented in the computer (when using the compiler relevant here) is approximately 10^{77} and the largest 'MAXINTEGER' is 10^9. If one tries to perform a computation which gives a result larger than this figure one will get a so-called 'arithmetic overflow' and the program run will be terminated.

Overflow can be avoided by excluding one or more terms in 'H-approx.' for critical values of n. However, one has to make sure that the excluded terms are so small that they are smaller than the precision of the computer. This can be done by a rough estimation, and does not need to be exact, since the limit for overflow is at 77 figures (compared to 16 decimals which is the precision of the computer).

- User instructions should be kept to a minimum (for example, 'Give a positive integer').
- The program should compute and print the sum for the value of n which is read in. ('The sum is: = ' is sufficient as a print out and the decimals should not be limited.)
- The program should ask for and read values of n from the terminal an arbitrary number of times.
- If the n-value 0 is given, the program should interpret this as the end of the indata and terminate the run.

References

Adelson, B., 1984. When novices surpass experts: The difficulty of a task may increase with expertise. Journal of Experimental Psychology: Learning, Memory and Cognition 10, 483–495.

Allwood, C.M., 1986. Novices on the computer: A review of the literature. International Journal of Man–Machine Studies 25, 633–658.

Allwood, C.M. and C.-G. Björhag, 1990. Novices' debugging when programming in Pascal. International Journal of Man–Machine Studies 33, 707–724.

Carver, S.M. and S.C. Risinger, 1987. 'Improving children's debugging skills'. In: G. Olson, S. Sheppard and E. Soloway (eds.), Empirical studies of programmers: Second Workshop. Norwood, NJ: Ablex, pp. 147–171.

Davis, R., 1983. User error or computer error? Observations on a statistics package. International Journal of Man–Machine Studies 19, 359–376.

Gilmore, D.J. and H.T. Smith, 1984. An investigation of the utility of flowcharts during computer program debugging. International Journal of Man–Machine Studies 20, 357–372.

Gray, W.D. and J.R. Anderson, 1987. 'Change-episodes in coding: When and how do programmers change their code?' In: G. Olson, S. Sheppard and E. Soloway (eds.), Empirical studies of programmers: Second workshop. Norwood, NJ: Ablex. pp. 185–197.

Gugerty, L. and G.M. Olson, 1986. 'Debugging by skilled and novice programmers'. In: M. Mantei and P. Orbeton (eds.), Proceedings CHI '86: Human factors in computing systems. Boston, MA: ACM, pp. 171–174.

Katz, I.R. and J.R. Anderson, 1988. Debugging: An analysis of bug-location strategies. Human–Computer Interaction 3, 351–399.

Miller, L.A., 1974. Programming by non-programmers. International Journal of Man–Machine Studies 6, 237–260.

Myers, G.J., 1978. A controlled experiment in program testing and code walkthroughs/inspections. Communications of the ACM 21, 760–768.

Nanja, M. and C.R. Cook, 1987. 'An analysis of the on-line debugging process'. In: G. Olson, S. Sheppard, and E. Soloway (eds.), Empirical studies of programmers: Second workshop. Norwood, NJ: Ablex. pp. 172–184.

Vessey, I., 1985. Expertise in debugging computer programs: A process analysis. International Journal of Man–Machine Studies 23, 459–494.

Waddington, R., and R. Henry, 1990. 'Expert programmers re-establish intentions when debugging another programmer's program'. In: D. Diaper, D. Gilmore, G. Cockton and B. Shackel (eds.), Proceedings of the Third IFIP Conference on Human–Computer Interaction – Interact '90. Amsterdam: North-Holland. pp. 965–970.

Wiedenbeck, S., 1985. Novice/expert differences in programming skills. International Journal of Man–Machine Studies 23, 383–390.

Acta Psychologica 78 (1991) 151–172
North-Holland

Models of debugging *

David J. Gilmore

Nottingham University, Nottingham, UK

This paper proposes a view of computer program debugging, which tackles some of the simplifying short-comings of existing models. The paper begins by reviewing some of the existing models of debugging and their assumptions, before looking in more detail at one of the dominant paradigms for investigating debugging, that of predicting bug detection success. A reanalysis of the bug detection data from Gilmore and Green (1988) provides evidence that the assumptions of existing models are not valid. The important part of this result is the realisation that these assumptions have been derived from a view of debugging as fault diagnosis, rather than as a critical component of design. In conclusion, the paper describes the important features of debugging as a design activity, before outlining some predictions and implications which can be derived from the model.

Introduction

To many it may seem surprising to ask the question 'why study debugging?', since it is obviously a central skill of computer programming, especially if maintenance is included within the definition of debugging. Yet, despite the fact that debugging is a skilled combination of both program understanding and program coding, it has surprisingly been ignored by many of those studying psychological factors in programming. An examination of Curtis's comprehensive review of different paradigms within the psychology of programming (Curtis 1988) reveals no mention of debugging.

Although it could be argued that debugging is simply a combination of comprehension and coding and, therefore, does not deserve a sep-

---

* The ideas and data reported here were presented first at the Second Psychology of Programming Interest Group Workshop (Walsall, UK) and I am grateful to the participants of that meeting, and to members of the Ergonomics Unit at University College, London for their useful comments, suggestions and criticisms.

Author's address: D.J. Gilmore, Psychology Dept. Nottingham University, Nottingham NG7 2RD, UK; E-mail: dg@psyc.nott.ac.uk.

arate research effort, this argument is not persuasive for those concerned with software reliability and the provision of debugging tools for professional programmers. Improved comprehension and coding tools may contribute to improved debugging, but without an understanding of the process of debugging itself, they cannot be the answer.

A number of papers on debugging do exist, as do a few models of debugging processes. However, many of these only provide an analysis of the act of bug location, without discussing the wider task context, which might include difficulties understanding the problem specification, or verifying that the algorithm does indeed solve the problem. Given the importance of debugging activities, it is important to clarify the validity of the conclusions drawn in these studies, since they could easily lead to the development of debugging tools which do not help, or which do not address the real problem. Also, many of the studies which have investigated debugging have done so simply as a means to investigate the cognitive processes of comprehension (e.g. Gilmore and Green 1988). In such studies it is not been considered necessary to articulate the relationship between debugging and comprehension.

Fig. 1 shows, schematically, a common approach to the context in which debugging is assumed to occur. In this approach debugging is taken to be comprehension, followed by bug location followed by bug repair. Bug location is assumed to be a process of inspecting a mental representation of the program code (the output of the comprehension process), which has been derived from the fixed entities – the code, the problem statement and real-world knowledge. Surprisingly, none of these is considered to have a direct impact on the bug location process, since their effect is mediated by comprehension. Fig. 2 gives an example of a particular version of this model, from Katz and Anderson (1988). The schematic model, which can be termed the *fault diagnosis* model, embodies three assumptions, each of which will be discussed briefly in the following sections:

(a) Comprehension always precedes debugging
(b) There is a single bug location process which is to be observed.
(c) The problem is not a problem, except insofar as it affects comprehension.

However, before proceeding it is necessary to clarify some of the terminology to be used in this paper. Comprehension is a difficult word since it is used to refer to both the act of comprehending and to the

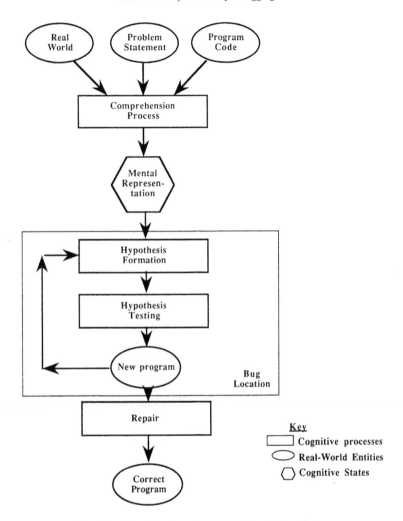

Fig. 1. A schematic view of many models of debugging.

resultant mental state of understanding. Thus, this paper will make a distinction between the *comprehension process* and the *state of comprehension*. The former involves the mobilisation of cognitive resources and processes in some particular configuration, with the goal of constructing some mental representation of the program code. It is this mental representation of the code which is the comprehension state. The possibility that the cognitive processes may be used in different ways in order to achieve a comprehension state gives rise to the

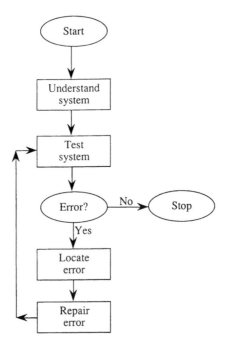

Fig. 2. A model of debugging (from Katz and Anderson 1988: 353).

importance of comprehension strategies. The choice of comprehension strategy may be determined by either process or state considerations, but the effect of that choice is to affect the comprehension process, and maybe indirectly the comprehension state.

Assumptions of debugging models

Comprehension precedes debugging

The task of debugging has been used to provide a window on the nature of the programmer's mental representation of the code. In such situations conclusions can only be drawn about the comprehension state if one assumes that the comprehension process is complete before debugging occurs. An example of a study whose conclusions require this assumption is that by Gugerty and Olson (1986), who conducted observational studies of novices and experts debugging programs written both in LOGO and in Pascal.

They found that experts were quicker and more successful than novices, and that they tested fewer hypotheses about the code, whilst novices frequently added more bugs to the program. This difference was attributed to the experts comprehending the program more successfully than did the novices. However, Gugerty and Olson observed that prior to forming any hypotheses, novices and experts engaged in similar proportions of activities such as reading the code and reading the problem statement (though experts were faster). From this they conclude that experts and novices use the same comprehension processes, but that experts are more successful and acquire a better mental representation of the code. The success of the experts at debugging is not attributed to better debugging skills, but to better comprehension, which is then followed by the application of the same debugging processes, but in relation to a better knowledge base for the experts than for the novices.

Thus, although Gugerty and Olson conclude that the differences in debugging success are due to differences in comprehension ability of two groups this depends upon the assumption that the process of comprehension is the same for both groups and that this process precedes debugging.

An explicit example of the assumption that the comprehension process is separate and that the comprehension state is complete before debugging begins is the study of LISP debugging by Katz and Anderson (1988), who write that

'when debugging, one must
   a) test the program....
   b) locate the erroneous line....
   c) rewrite the buggy code.
Also, if the person did not write the code originally (or wrote it a long time ago), that person
will probably need to *first comprehend* the program.' (ibid, p. 353, my emphasis).

A single debugging process

Alongside the assumption that comprehension is separate and comes first, it is also commonly assumed that there is only one debugging process. Given this assumption we can infer that when we observe differences in debugging success, we can conclude that they are due to differences elsewhere – either in comprehension or knowledge structures for instance.

An interesting example of work which assumes that there ought only to be one debugging process is Kessler and Anderson (1986), who provide a four-stage model of novice LISP debugging; the stages being code comprehension, bug detection, bug localization and bug repair. Kessler and Anderson collected protocols from novice LISP programmers and claim a good fit between their model of debugging and the protocols. However, this ignores the one subject who generated the correct code mentally and compared this with the actual code. Kessler and Anderson disregard this subject arguing that this strategy could only work for very simple problems. The fact that their model only allowed one debugging process prevented them appreciating the importance of this non-comprehension-based exception.

Another example of work assuming a single debugging process is that of Iris Vessey. Vessey (1985) proposed a model which assumed that novices and experts engage in the same process, but with different knowledge structures available to them. Her analysis of debugging protocols showed novices flitting between different stages in the model of debugging, whilst experts moved between stages quite rarely. Thus, the experts performed smoothly whilst the novices were erratic and error-prone. However, this conclusion depends upon the assumption that the novices were trying to perform the same process as the experts. If they were actually using a quite different debugging strategy then they might appear to be behaving erratically, when in fact they were competently pursuing an inefficient and error-prone strategy.

Vessey (1989) adopts the model of debugging used by Atwood and Ramsey (1978) in one of the first studies of debugging. This model is derived from Kintsch's theory of text comprehension, which is based on an analysis of the propositions contained within the text. During comprehension the propositions are chained together by links in the operands to form the programs microstructure. The meaning can then be extracted by reforming this into a macrostructure, which represents the semantics of the program. The mental representation of a program is assumed to be both the micro- and macro-structures. Both structures are arranged hierarchically.

Such a model predicts that processing information which is high in the hierarchy will be easier than processing that which is deep down in the structure, except during the comprehension process, when the lower details may be processed first. Thus, Vessey predicted that bugs which are high in the propositional structure should be detected more easily than bugs deep in the hierarchy. Note that this argument requires assumption (a): that the comprehension process is complete before debugging occurs.

These arguments do not inform us how the bugs are detected, simply which bugs should be detected most easily. Without a process model it is difficult to be certain, but Vessey's arguments would seem to imply that the debugging process is the same for all contexts, since depth in the propositional structure is presented as the key factor affecting the detection rate.

However, Vessey's data provided little support for her model and she suggests that widely differing debugging strategies and processes may have been used, though other interpretations are possible. The main weakness in the conclusions is that they are based on an analysis of only one bug, occurring in different places. Not only does this lead to problems of generalization, but it is not clear that the same bugs occurring in different locations can really be called 'the same'. Control-flow bugs may seem to be easily equated, but semantic and plan-bugs cannot be equivalent unless at the same point in the program. For this reason it is not possible to conclude that the level of the bug in the two structures studied by Vessey was the only difference between the programs used. One large difference, mentioned but not discussed by Vessey is the fact that the location of the bug had different effects on the output. This difference was so great that even after considerable study, some subjects in one of the four conditions denied there was any difference between the actual output and that desired.

The potential for external factors such as the output, or maybe the problem statement to have an effect at the debugging level is something which is omitted from the schematic view described above. But there is some evidence which suggests that the programmer's understanding of the actual problem may be important.

The problem statement

Katz and Anderson (1988) describe a large number of LISP errors and look at the frequency with which novices repeat similar errors across a variety of problems. Regardless of any problems associated

with identifying the 'same' errors across different contexts, their conclusions critically depend upon the error generation and detection
processes being independent of the actual problem.

Thus, because the 'same operations' are performed erroneously in
some contexts, but correctly in others, they conclude that the novices
are only making 'slips' rather than applying misconceptions about how
LISP works. However, the 'same operations' are defined according to
the correct implementation of the program and, thus, students who
misunderstand the problem will generate an error, which is neither a
slip nor a misconception about LISP. This can be seen in some of the
example errors provided by Katz and Anderson. For example, one
error described occurs when the programmer writes code specific to the
example data provided in the problem description rather than general
purpose code.

The distinction made by Wertz (1982) between *conceptual* and
teleological bugs is useful here, since it clarifies the role of the problem
in understanding bug generation and detection. Conceptual bugs are
bugs arising from discrepancies between actual behaviour and required
behaviour (cf. specification), whereas teleological bugs are discrepancies between actual behaviour and that intended by the programmer. In
the latter case the program contains a *traditional* error, but in the
former the program may be syntactically and semantically correct, the
error arising because the program solves the wrong problem.

If programming behaviour reliably generated *either* conceptual *or*
teleological bugs then understanding debugging might not be too
complicated, but the crux of the problem is the fact that they can
co-occur. A simple conceptual error, compounded by a teleological
error may lead to a program whose behaviour is extremely complex and
whose repair is very difficult. A further important point made by
Waddington and Henry (1990) is that, for programmers who didn't
write the original program, all bugs are conceptual, until inferences
have been drawn about the intentions of the original programmer. This
reverse process, from code to problem statement, is often omitted from
models of program comprehension and debugging.

Summary

The models of debugging commonly used are simplistic and tend to
rely upon a distinction between comprehension and debugging. In

general the research takes a *fault-diagnosis* approach to debugging, in which one simply detects a discrepancy between *what should be* and *what is*, before correcting it. To enable the study of debugging separately from comprehension and coding it is necessary to make these three assumptions about the nature and ordering of comprehension and debugging.

This fault-diagnosis view of debugging implies a static model of comparison between the code and a mental representation. The contrast can be made with a design-oriented view of debugging in which a dynamic comparison occurs between static code and a dynamic mental representation of a dynamic process.

This possibility is tentatively suggested by Iris Vessey (1989) at the end of her inconclusive paper, where she suggests that debugging might depend upon the formation of an appropriate mental model of correct program functioning. A question which she raises is whether 'a control-flow bug, for example, requires the formulation of a mental model that foregrounds the control-flow of the program?' (p. 45). It is this possibility which is investigated in the following analysis of some bug detection data. Anticipating the results, it is fair to say that different mental models are required for different bugs. A model of debugging is required in which the debugging and comprehension processes are much more closely connected.

An analysis of bug detection data

Introduction

With these considerations in mind, and particularly given some of the problems arising from Vessey's use of a single bug, I re-analysed the data from the experiment by Gilmore and Green (1988), in which 80 subjects debugged 10 versions of each of two programs, with each program containing 2 bugs. Forty subjects were Pascal programmers and debugged programs written in Pascal and forty were BASIC programmers who debugged equivalent programs (with equivalent errors) written in BASIC. Obviously it is impossible to construct identical programs with identical errors in two different languages, and in some cases the errors could not be equated and it was necessary to introduce different but comparable errors into the BASIC programs. Further information on the programs can be found in Gilmore and Green (1988) and Gilmore (1986).

In the original analysis the emphasis was on experimental manipulations of presentation format and no attention was paid to the bugs themselves. In the analysis to be reported here, the opposite is true, since here I shall analyse the data by bugs, rather than by subject groups. The analysis to be described will apply multiple regression techniques, using bug detection rate as the dependent variable.

If debugging can be characterised as a static, fault diagnosis process, then we would expect to be able to predict detection rates on the basis of *depth* within the comprehension structure. If, however, debugging is a dynamic, design-like activity then we would expect bug detection rates to be specific to their particular context and the programmer's chosen strategy.

The programs and errors

The programs used in the study were all attempts at the classic 'rainfall' problem used in numerous Yale studies of programming skills. The bugs introduced into these programs were adapted from the Yale Bug Catalogue (Johnson et al. 1983), but were classified into four types (giving rise to three dummy variables in the regression):

- *surface*: typing errors and similar mistakes;
- *control*: errors in the control structure, which could be termed syntactic (omissions or ill-formed statements);
- *plan*: errors in the plan structure of the program, where the surrounding control structures are correct (omitted update of count variable);
- *interactions*: errors which arise from a combination of control errors and plan errors (e.g. Initialisations in the wrong place).

This classification was chosen for its correspondence with the presentation manipulation and for a behavioural, rather than symptomatic description (see Gilmore and Green (1988) for more information). Further experiments (Davies 1990) have supported the usefulness of this classification, though I would not wish to claim that this is the only possible classification of these errors into different types. [1]

Measures of bug depth

For the reanalysis, a measure of the depth of each bug was necessary. In order to avoid the analysis making presumptions about the nature or timing of the comprehension process it was necessary to consider more possible models of the comprehension process than the Kintschian model considered by Vessey. I, therefore, adopted three measures of depth, or location (i.e. three more independent variables), each representing depth within a different structure:

(1) *Serial structure*: As the name implies this was simply the serial position of the bug in the code. It was measured by taking the line number of the bug. Since the two

---

[1] Other differences between these bugs, such as their effect on the output of the program (e.g. runtime errors or simply an incorrect result) or the distinctiveness of the resulting statement (e.g. omissions or the use of the wrong operator in an assignment) have not yet been addressed, even though they may be major contributors to bug detection rate.

programs were of comparable length, no adjustment was made for program length (i.e. the proportional position in the code). A hand-simulation strategy for comprehension might lead to this being the critical structure, with bugs nearer the beginning of the program being more easily located.

(2) *Propositional structure*: This term is used in the sense used by Vessey, where it referred simply to the depth of the bug in the control structure of the program. It should be noted that this is not analogous to Kintsch's propositional structure which captures the importance of each proposition to the meaning of the text. Propositional depth was measured by counting $+1$ for each new control structure encountered and -1 for the end of each control structure. A control structure dominated model of comprehension (top-down or bottom-up) would predict that this depth will be critical.

(3) *Goal structure*: This is closer to Kintsch's propositional analysis and is based on the Yale analysis of programming goals and plan structures. The analysis was performed with reference to the Yale goal structures for comparable problems, but it was not exactly equivalent. Fig. 3 gives a graphical representation of the goal structure for one of the two problems, indicating the depth of various subgoals. A plan-based model of comprehension would predict that bugs occurring higher in these structrues should be detected more easily.

In order to ensure that these are three different structures, the correlations between the depth of each errors in each pair of structures was calculated. The largest correlation occurred, somewhat surprisingly between depth in the serial structure and depth in the propositional structure ($r = 0.24$), which is small enough not to give cause for concern.

These measures were calculated separately for every program and every bug. This means that there is not complete equivalence on these measures between the Pascal and BASIC programs. However, the range of depths is the same for each language.

Task and method

In the experiment subjects were given 10 buggy versions of programs for each of two problems. All subjects received programs in their language of expertise. Each buggy program contained 2 bugs, giving a total of 40 bugs overall, ten of each of the four types. The subjects (40 experienced Pascal programmers and 40 experienced BASIC programmers) were given a sample correct program for each problem, in an attempt to ensure that they fully understood the problem which the buggy programs were supposed to solve. All versions of the program varied in terms of variable naming, details of the algorithm used, etc. ensuring that the task was not a simple pattern matching one. Furthermore, the sample correct programs were not available for inspection whilst the buggy programs were being inspected. Subjects were told that their task was to mark and briefly describe any errors which they detected in each of the 10 programs for each problem, but that they were not to try and correct the program. They were given limited time to study each program and the analysis here only examines detection rate (ie the number of subjects who detected the bug), and not the time to detection.

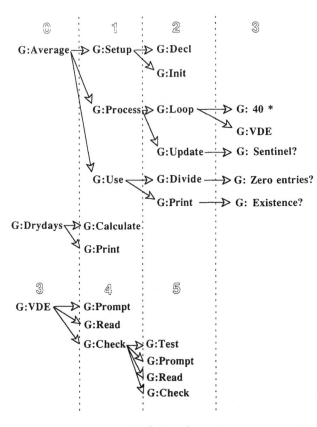

Fig. 3. The goal structure of problem 2, indicating the depth measure for various subgoals.

Results

Pascal programmers

Multiple regression was performed on the data, using detection rate as the dependent variable and using bug type and bug depth as 6 independent variables. This complete model (Model a) accounted for 36% of the variance ($F_{6,33} = 4.69$, $p = 0.001$). When the variables coding bug type were dropped from the model (Model b), this changed to 5% of the variance (a significant change, $F_{3,36} = 6.77$, $p = 0.001$). By contrast, when the depth measures were omitted, the model (Model c) accounted for 37% of the variance (a non-significant change). This analysis is summarised in table 1, from which one can infer that bug type is the most important factor influencing detection, with the location of the bug in the program making no contribution.

However, given Gilmore and Green's original results concerning the improved detection of bugs when perceptual cues highlighted the relevant structure it seems sensible to examine regression models which include interaction terms (i.e. the product

Table 1
Multiple regression summary for the Pascal programmers (main effects only).

Model	R^2	Adj. R^2	F	p	F ch.	p ch.
a: 6 IVs	0.46	0.36	4.69	0.001		
b: 3 IVs – bug depth	0.13	0.05	1.76	0.172	6.77	0.001
c: 3 IVs – bug type	0.42	0.37	8.62	0.000	0.86	0.47

of location and bug type for each location and each bug type). Thus, a second analysis was performed for the purpose of clarifying the role of depth for each bug type separately. For this analysis, there were 12 predictors in all – the three original measures of depth, plus the nine interaction terms (three depth variables times three type variables). This model (Model d) was able to account for 49% of the variance ($F_{12,27} = 4.1$, $p = 0.001$). From this model the three critical interactions (goal depth × plan bugs, propositional depth × interaction bugs and serial position × control flow bugs) were omitted, producing a model (e) which accounted for 23% of the variance, which is a significant change ($F_{3,30} = 5.98$, $p < 0.005$). Although Model e produced a significant regression equation, none of the individual coefficients had a significance less than 0.1. When the other 6 interaction terms were omitted instead, the model (f) was still able to account for 44% of the variance (a non-significant change – $F_{6,33} = 1.5$, $p = 0.2$). These results are summarised in table 2.

In Model f the coefficients for each of the three interaction terms were positive, indicating that the bugs were easier to find when they were deeper in the structures, which is in contradiction with the usual predictions in such experiments.

BASIC programmers

Since the original results for the BASIC programmers were very different from those for the Pascal programmers we did not expect to see the same picture here and stepwise regression was used initially to uncover the pattern of results. Table 3 shows the summary table for the multiple regression with just the 6 independent variables (interaction terms excluded). From this it is clear that the probability of bug detection

Table 2
Multiple regression summary for the Pascal programmers (including interaction terms).

Model	R^2	Adj. R^2	F	p	F ch.	p ch.
d: 12 IVs	0.65	0.49	4.1	0.001		
e: 9 IVs – bug depth, 6 interaction terms	0.41	0.23	2.32	0.04	5.98	0.003
f: 6 IVs – depth, 3 interactions (goal × plan, prop × intern, serial × control)	0.52	0.44	6.06	0.000	1.54	0.2

Table 3
Multiple regression summary for the BASIC programmers (main effects only).

Model	R^2	Adj. R^2	F	p	F ch.	p ch.
a: 6 IVs	0.3	0.17	2.33	0.05		
b: 5 IVs – excl. propnl. depth	0.13	0	0.97	0.5	8.1	0.008
c: 1 IV – propnl. depth	0.2	0.17	8.9	0.005	1	0.43

in these BASIC programs is predictable from the depth of a bug in the propositional (control-flow) structure, though the proportion of variance explained is substantially smaller than in the Pascal analysis. It is also worth noting that the coefficient for propositional depth is positive, indicating that bugs were more likely to be found, the deeper they were in the structure.

If the interaction terms are included in the analysis (see table 4) we find exactly the same result, namely that only depth in the propositional structure offers any prediction of bug detection success.

Summary

These regressions results provide a clear indication of the importance for Pascal programmers of bug type on debugging success (accounting for 36% of the variance). The results of the second regression, with the interactions included, are striking in that they suggest that the detection rate for different bugs may differ according to their depth in one particular structure of the program. If the psychological theorizing about the role that depth plays is correct then this implies that Pascal programmers (as a group, at least) must be capable of mentally representing these different structures.

The BASIC programmers, who lacked explicit instruction in program design (cf. Davies 1990), seemed to have a clear preference for comprehension via the control-flow structure and propositional depth thus accounted for nearly 20% of the variance in their bug detection success. This result is not too surprising and it confirms the original analysis of these data, which showed how BASIC programmers described most of the errors detected in terms of corrections to the control flow. [2]

The fact that, in both cases, bugs were more easily found when they were deeper in the structures contradicts the traditional interpretation of propositional analyses of comprehension. Both groups of programmers were more likely to find the deeper bugs, suggesting that they were being detected *during* an ongoing comprehension process, rather than by comparison with a comprehension state.

Both of these points challenge the assumptions about debugging which were described earlier, since our data indicate that:

[2] Gilmore and Green (1988) discuss the various reasons why these language/experience differences may have occurred. However, these differences are not pertinent to the current discussion.

Table 4
Multiple regression summary for the BASIC programmers (including interaction terms).

Model	R^2	Adj. R^2	F	p	F ch.	p ch.
d: 12 IVs	0.44	0.19	1.76	0.1		
e: 11 IVs – excl. propnl. depth	0.23	− 0.1	0.77	0.67	10	0.004
f: 1 IV – propnl. depth	0.19	0.17	8.9	0.005	1.1	0.4

- debugging seems to be occurring as a part of an ongoing, comprehension process;
- for at least the Pascal programmers there seem to be different debugging (or comprehension) strategies available, each focussing on different structures;
- although bug type is a critical factor, the programmer's knowledge of the problem can also have an impact on some of the types of bug.

Most importantly, these new analyses indicate that comprehend-and-debug explanations of debugging are not adequate. Although it may be possible to use such models to explain the data from our BASIC programmers, they are wholly inadequate for the richness of the Pascal programmers' performance.

A model of debugging

What the above analyses show most clearly is the complexity of bug detection, but in particular they illustrate the fact that debugging success is a product of both comprehension processes and the actual meaning of the task, program and bug. In order to accommodate these factors we need a more flexible, strategic model of debugging behaviour. This will have three components.

(1) A flexible, incomplete comprehension process that can choose to represent different program structures according to task demands, tools available and skill level.
(2) A mental representation of some aspects of a 'correct' program for the problem (e.g. behaviour, structure, etc.) as well as of the buggy program. [3]
(3) A process for detecting mismatches across these two representations of the problem and program.

---

[3] This is seen in our experiences of teaching programming to undergraduate psychologists, where the frequent stumbling block is not understanding the language and its syntax, but understanding the problem which is to be solved.

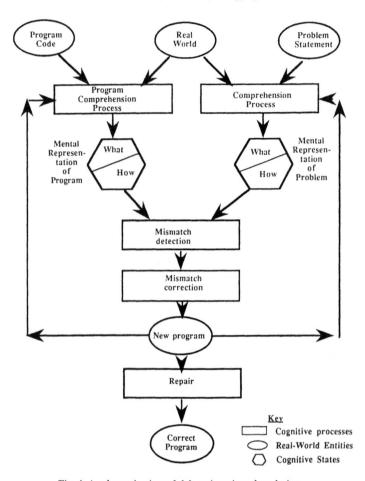

Fig. 4. A schematic view of debugging viewed as design.

Such a model can be described as a reconstructive model, in which the programmer debugs through reconstructing some of the original design decisions and verifying them against the actual performance of the program. This model is offered in contrast to the sequential comprehend and debug model. It should be noted, however, that the model described is capable of generating many debugging strategies, including comprehend-and-debug behaviour.

Fig. 4 illustrates the schematic view of the proposed model, in which comprehension clearly plays a central role. One important goal of the comprehension process is to bring the two representations (of the buggy and of the correct program) into a common form which can be easily compared.

Comprehension

The important characteristic of comprehension as presented here is an emphasis on both the program's behaviour and the code: i.e. both *what* it does and *how* it achieves it. In order to comprehend a program fully the programmer should have knowledge of both *what* and *how*, but in order to perform numerous programming tasks (including much maintenance and debugging) only knowledge of one of these is really necessary. A variety of factors may determine which structure the programmer focuses on. For example, the programming environment may provide better facilities for observing the behaviour of the program (*what*), than the notation does for teasing apart how the program actually works (*how*). BASIC would be a good example of such a language. Although the programmers in the experiment described did not have access to any computational environment, their experience of BASIC may have promoted a comprehension strategy based on *what* information.

A complexity here arises from the recursivity of the programming process, in which one can treat parts of the program as programs themselves. Indeed, this is exactly the point made by Koeneman and Robertson (1991) who demonstrated that the comprehension of large programs is achieved by focussing in on small manageable parts of the whole, thus reducing the problem to one of comprehending a small sub-program. This leads to the characteristic that the mental representation of the program can contain many different levels. How many levels will depend upon the size of the programming problem, the skill level of the programmer (novices may have difficulty comprehending even single statements – cf. Mayer 1979), the familiarity of the problem and the type of mental representation being formed. In this way certain parts of the program can be represented by *what* knowledge and other parts by *how* knowledge.

The distinction between *what* and *how* information requires clarification. *What-it-does* information is that information which can be extracted and ordered from the individual program statements, whereas *how-it-works* information involves extracting the relationship between the program and the computational architecture.

In most programming environments it is *what* information which is most easily obtained, but the difference between *what* and *how* knowledge is affected by the language paradigm. For a declarative language (e.g. Prolog) procedural information would be *how-it-works* knowledge, whilst for BASIC, such procedural information would represent *what* knowledge. In this way one can see why tracing facilities are critical in a language like Prolog, since they provide the easiest access to *how* information about a Prolog program, whereas Pascal and BASIC programmers prefer breakpointing and variable tracing in order to acquire *how* information.

Debugging

Most programmers may feel satisfied with their comprehension when they understand either the *what* or the *how* information. However, when debugging the programmer is not completely free to choose which information to gather, since they need to be able to compare their comprehension model with their model of what or how the correct program should be.

We can assume that there can be as much complexity surrounding the comprehension of the problem as there is for the buggy program, but there is no reason (especially when debugging someone else's program) for the two comprehension models to be of the same form. However, when searching for a mismatch between problem and program (i.e. debugging), it is essential for there to be some overlap between the models.

There are three important features of this theory of debugging:

(1) Programmers must first reconcile their understanding of both program and problem.
(2) Mismatch detection can take place in terms of either *what* or *how* information, dependent upon the nature of the programmer's knowledge of program and problem, and the support provided to them by the environment.
(3) Mismatch correction can be performed in relation to both *what* and *how* information.

Thus, if a programmer comprehends a buggy program in terms of *what* it does, then debugging should be easiest if their understanding of the problem is in similar terms. An important property of this model is that it captures the potential for conflict between the static representation of the problem solution (the program code) and the dynamic solution (program behaviour). It is this characteristic which leads to debugging as a design activity, since it involves a study of how the detailed dynamics of the algorithm map onto static code.

The one task which is not fully described within this framework is that of mismatch detection, which is, of course, the crux of debugging behaviour. However, given this model of debugging we are in a much better position to conduct research into how differences between the problem and the program are detected.

Implications

The first important implication of this theory of debugging is that it characterises debugging as design, rather than as fault diagnosis. The model outlined in fig. 4 describes programming behaviour in general as well as it does debugging behaviour. The only difference is the quantity of information derived from the program code and the problem statement. One can, thus, view programming as a continuum from early design and coding (all information from problem statement) through debugging one's own program (equal information from program code and mental representation of problem) through to debugging someone else's code (equal information from program code and problem statement).

Although I have previously written about the importance of a distinction between coding and comprehension and, indeed between different types of comprehension, preferring to distance myself from the 'comprehension is the core of all programming' school of thought (e.g. Gilmore and Smith 1984), the view being described here does not contradict this. The danger of describing comprehension as the core programming process is that it is assumed to be a simple unitary process, rather than a complex strategic one. In the model presented here comprehension is central process, but subject to strategic variations.

The task of understanding *how* a program works can be very demanding, especially for novice programmers (e.g. due to working memory limitations), who find it easier to focus on *what* knowledge.

This leads them to a process of executing the program to see what it does, followed by a comparison with their mental model of the desired behaviour. Changes are then made to the program in order to make the actual and the desired behaviour more similar. In this way they may continually introduce new bugs into programs, since they are simply attempting to alter the behaviour of the program, rather than trying to correct errors in the conception, or construction of the algorithm.

As well as providing this explanation of why novices introduce many new bugs to their programs, the model also provides a new perspective on the results of Davies (1990). He showed that BASIC programmers without education in program design skills did not use programming plans to aid their comprehension and debugging processes, whilst BASIC programmers who had been taught about program design performed like Pascal programmers, being guided by the plan structure of the program. Gilmore and Green (1988) had observed the behaviour of the first group and explained this in terms of notational differences (using the concept of *role-expressiveness*) which gave rise to different knowledge structures. Davies rejected this notational difference, claiming that design skills were a contributory cause of the Pascal–BASIC differences observed by Gilmore and Green. Within the new framework presented here, it is possible to see how both the notation and the existence of design skills could give rise to strategic differences amongst programmers, leading some groups to focus on *how* information, comparable to programming plans, whilst others focus on *what* information (through control-flow).

The importance of congruence between the programmer's models of the program and the problem may also explain the results of Petre and Winder (1988) who found that, despite frequent arguments that different languages are suited to different kinds of problems, skilled programmers used the same algorithm for the same problem across a wide range of programming languages. This finding does not refute the claim that different languages are suited to different problems, particularly since only half the experts used the same language first for all the presented problems, but it does indicate that the cognitive cost of reconstructing one's understanding of the problem is expensive relative to the cost of translating it into a different language.

From the system designer's perspective an important implication in these results and this model of debugging is that there is a clear, critical role for the programming environment in supporting succesful debugging performance, something which is remarkably lacking from earlier models. Environments which provide 'good' sequential trace information for procedural languages may be reinforcing a *what-it-does* comprehension strategy, when *how-it-works* knowledge might be more important to the programmer whose understanding of the problem is of this form. Paradoxically tools which support the programmer in finding an alternative perspective on the code may be viewed as irrelevant by a programmer who does not realise that an alternative exists and whose model of the problem is congruent with their preferred comprehension strategy.

Summary

From the review of existing models of debugging and from the regression analyses on the detection rates of various errors, I have developed a model of debugging which places it amidst normal programming activities (i.e. as a part of design), rather than separating it out as a different activity (i.e. fault diagnosis). One of the model's main features is the important role given to the mental representation of the problem to be programmed, as well as the mental representation of the program itself. The act of debugging is one of reducing the difference between the actual code and that required, through a comparison of these two mental representations.

References

Atwood, M.E. and H.R. Ramsey, 1978. Cognitive structures in the comprehension and memory of computer programs: An investigation of computer program debugging. NTIS, AD, A060522/0.

Curtis, W., 1988. 'Five paradigms in the psychology of programming'. In: M. Helander (ed.), Handbook of human–computer interaction. Amsterdam: North-Holland.

Davies, S., 1990. The nature and development of programming plans. International Journal of Man–Machine Studies 32, 461–481.

Gilmore, D.J., 1986. The perceptual cuing of the structure of computer programs. Unpublished Ph.D. thesis, University of Sheffield.

Gilmore, D.J. and T.R.G. Green, 1984. Comprehension and recall of miniature programming languages. International Journal of Man–Machine Studies 21, 31–48.

Gilmore, D.J. and T.R.G. Green, 1988. Programming plans and programming expertise. Quarterly Journal of Experimental Psychology 40A, 423–442.

Gilmore, D.J. and H.T. Smith, 1984. An investigation of the utility of flowcharts during computer program debugging. International Journal of Man–Machine Studies 20, 357–372.

Gugerty, L. and G. Olson, 1986. 'Comprehension differences in debugging by skilled and novice programmers'. In: E. Soloway and S. Iyengar (eds.), Empirical studies of programmers. Norwood, NJ: Ablex.

Johnson, W.L., E. Soloway, B. Cutler and S. Draper, 1983. Bug catalogue: I. Research Report No. 286, Computer Science Dept, Yale University, Newhaven, CT.

Katz, I. and J.R. Anderson, 1988. Debugging: An analysis of bug location strategies. Human–Computer Interaction 3, 351–400.

Kessler, C.M. and J.R. Anderson, 1986. 'A model of novice debugging in LISP'. In: E. Soloway and S. Iyengar (eds.), Empirical studies of programmers. Norwood, NJ: Ablex.

Koeneman, J. and S.P. Robertson, 1991. 'Expert problem-solving strategies for program comprehension'. In: S.P. Robertson, G. Olson and J. Olson (eds.), Proceedings of CHI '91. New York: ACM Press. pp. 125–130.

Mayer, R.E., 1979. A psychology of learning BASIC. Communications of the ACM 22, 589–593.

Petre, M. and R. Winder, 1988. 'Issues governing the suitability of programming languages for programming tasks'. In: D. Jones and R. Winder (eds.), People and computers IV. Cambridge: Cambridge University Press.

Vessey, I., 1985. Expertise in debugging computer programs: A process analysis. International Journal of Man–Machine Studies 23, 459–494.

Vessey, I., 1989. Toward a theory of computer program bugs: An empirical test. International Journal of Man–Machine Studies 30, 23–46.

Waddington, R. and R. Henry, 1990. 'Expert programmers re-establish intentions when debugging another programmer's program'. In: D. Diaper, D. Gilmore, G. Cockton and B. Shackel (eds.), Proceedings of Interact '90. Amsterdam: North-Holland.

Wertz, H., 1982. Stereotyped program debugging: An aid for novice programmers. International Journal of Man–Machine Studies 16, 379–392.

Acta Psychologica 78 (1991) 173–197
North-Holland

Analogical software reuse

Empirical investigations of analogy-based reuse and software engineering practices

Alistair Sutcliffe * and Neil Maiden

City University, London, UK

This paper presents three empirical studies of software engineering behaviour during systems analytic and analogical reuse tasks. The motivation behind these studies was to inform the design of support tools for the early stages of software development. A first study investigated inexperienced software engineers during a systems analytic task and revealed that they encountered considerable difficulties. The second and third studies investigated analogical software reuse as one means of overcoming these difficulties. The second study involved an experimental investigation of the effectiveness of specification reuse on analytic performance. Reuse proved beneficial, although it appeared to lead to mental laziness manifest as specification copying. In contrast, a third study invesitgated analogical specification reuse by expert software engineers, to determine how successful analogical comprehension and reuse may best be achieved. Implications of findings from all three studies for the design of support tools during software development are reported.

Introduction

Cognitive studies of software engineering have focused on the solution of well-structured problems, for example the construction of a simple program in a specified programming language (e.g. Adelson 1984). However, software engineering involves two distinct phases; first acquisition and comprehension of a problem domain, and second creation of a software design representing a solution. Programming activity occurs during the second phase. The first phase, systems analysis, is more complex yet it is poorly understood. Some studies have identified a few determinants of good analytic performance (e.g.

---

* Requests for reprints should be sent to A. Sutcliffe, Dept. of Business Computing, School of Informatics, City University, Northampton Square, London, EC1V OHB, UK; E-mail: cc559@city.ac.uk.

Vitalari and Dickson 1983), however, no process model of the reasoning underlying analytic tasks has been developed.

Better planning and more critical testing of hypotheses have been suggested as some qualities which differentiate expert from novice analysts (e.g. Guindon and Curtis 1988). Furthermore, experts appear to use better heuristics and retrieve richer knowledge structures from memory. Psychological studies of program designers have demonstrated that novices have fewer preformed mental schema which can be retrieved from memory and that novices tend to focus on the surface aspects of the problem (i.e. lexical/syntactic features of the programming language) rather than the semantic level of the problem itself (e.g. Adelson 1984; Koulek et al. 1989). In addition, studies of program debugging have suggested novices fail to scope problems, resorting to a strategy of bug isolation and repair, whereas expert strategies are directed towards building multiple domain models. Systems analysis requires different and in many ways more demanding skills than programming, hence novice software engineers might be expected to have difficulty during an analytic task. A mental model of the analytic processes used by novice software engineers is necessary to inform the design of future CASE tools. This paper reports an investigation into an applied mental model of the analytic process for software engineering, based on protocol analyses of software engineering practices, followed by experimental studies on analogical reasoning during systems analysis.

Guindon and Curtis (1988) cite analogical reasoning and the reuse of domain knowledge as important determinants of expert analytic performance. Experts are able to retrieve previous knowledge stored in memory, however, novices do not have past experiences which are likely to help the analytic task. Software reuse is an important source of domain knowledge which novice software engineers could exploit by analogical reasoning. Many theories of analogy have been proposed, for example Gentner (1983), however, these theories are often incomplete and few task-related studies of analogical reasoning have been reported.

Analogous reasoning in problem solving is knowledge-intensive, hence problem solving by analogy is not suitable for complete automation. It is usually considered to have three steps, which mirror the steps involved in software reuse: (i) retrieval, (ii) transfer, and (iii) translation. In our scenario retrieval of reusable software is likely to be a tool-based task, however, transfer and translation of analogous software requires the software engineer. Two empirical studies are reported which investigated mental models of analogous reasoning in software reuse. One study investigated whether software reuse enhances analytic performance while other studies examined the mental processes of novice and expert software engineers during reuse, to enhance the previously described applied mental model of the analytic task. The problems presented to the subjects in the latter two studies were similar in scale to that provided in the first study, so that results from all 3 studies can be discussed in a similar context.

Study 1 – Applied mental model of the analytic process

An empirical study investigated cognitive processes of novice software engineers engaged in systems analysis. The objective was to describe the mental processes of systems analysis and create a model which had prescriptive value in that context, rather than a more general model of reasoning (e.g. GPS (Newell and Simon 1972) and ACT * (Anderson 1983)). Previous studies had investigated the behaviour of expert software engineers, hence this study investigated the analytic behaviour of novices, to contrast these previous studies and to provide an empirical basis for the diagnostic component of an intelligent CASE tool. A more detailed report of this study is given in Sutcliffe and Maiden (in press).

Method

Protocol analysis was used to investigate systems analysis behaviour of 13 novice software engineers (11M, 2F students in MSc in Business Systems Analysis, with a maximum of 6 months structured analysis experience). Three pilot subjects undertook the analytic task beforehand, to refine the problem and the experimental procedure. Subjects were asked to develop a specification for a delivery scheduling system described in narrative form. All subjects had background domain and method knowledge necessary to develop a specification. Domain knowledge included all aspects of a delivery scheduling problem, obtained by subjects from the narrative through experience on a case study. Method knowledge refers to procedural steps dictated by the Structured Systems Analysis (SSA) method, and its main representation technique, data flow diagrams (De Marco 1978). SSA is typical of techniques which currently exist to assist the software engineer during systems analysis. These techniques had recently been taught and practised as a part of the subject's MSc Curriculum.

All subjects were given a 400-word narrative describing the problems of a manual delivery scheduling system. The objectives for a computerised delivery system were also outlined, together with an example report describing the required delivery schedule. Subjects had no access to other materials during the protocols.

Subjects were requested to think aloud and their verbal protocols were recorded on audio tape. During the protocols subjects were advised to take their time, and not be afraid of verbalising too much, following the practice of Ericsson and Simon (1984). Instructions for subjects were read by the experimenter. Each subject was requested to develop a specification of a computerised system, using the SSA and data flow diagramming techniques taught on their MSc course.

Subjects were given 35 minutes to complete a solution, as pilot studies indicated this was sufficient time to complete the task. All subjects were expected to complete a solution by the end of it, and were halted after 35 minutes. Upon completion of the task subjects were retrospectively questioned by the experimenter about the analytic behaviour for a further 10 minutes. Retrospective questioning was driven by a checklist of different behaviours which were expected to occur during the task (e.g. 'Why did you model the current system before the required system?'). Finally all subjects completed a questionnaire identifying details of all previous analysis and programming experience.

Protocol transcripts were analysed by matching mental behaviours to speech segments, usually sentences or incomplete utterances. Six major categories were used, divided into four mental and two non-mental behaviours:

– *Recognise goal*: statement of a higher level problem goal, often a system requirement, used in structuring the problem space (e.g. '...And there is a need to improve the delivery system');
– *Assertions*: verbalisation of a belief or statement of facts about the problem domain, directly attributable to the problem narrative (e.g. 'Lorries leave the depot half empty');
– *Reasoning*: verbalisation of the creation, development and testing of hypotheses about the problem or its proposed solution (e.g. '...he also wants...to know about urgent orders, so again, that should be marked in some way, he will be able to pick it up straight away...');
– *Planning*: meta-level control over the analytic process (e.g. 'I'll list the inputs, outputs and sources, then draw a logical new DFD' or 'I'll read the problem once, then construct a specification').
– *Information acquisition*: searching for and retrieval of data from the problem text (e.g. reading the problem text);
– *Conceptual modelling*: physical construction of the system specification, recorded as a data flow diagram.

Reasoning was distinguished from Assertions by the degree of inference applied and concurrent non-mental behaviour (e.g. reading indicated assertions). Planning differed from Goal recognition in that it was a domain-independent statement of intention: goals state what the system must achieve whilst plans state how the subject develops a specification to meet those system goals.

The concurrent protocols were analysed to investigate reasoning behaviour in more detail. The life history of each hypothesis was traced by its thematic content until eventual rejection or resolution. Hypothesis reasoning behaviour was categorised as generate, develop, test, confirm, modify and discard.

Protocol categorisation was validated through cross-marking by two independent observers. Each observer allocated a behavioural category to each utterance in five randomly-selected protocols. Initial inter-observer agreement was 79% of the utterances categorised by both observers. Resulting differences were attributable to identifiable discrepancies between the observers (e.g. failure to recognise assertions made from the problem text). These differences between protocol categorisations were discussed and where necessary changes were agreed and reconciled by both observers.

Completeness scores were allocated to each subject's solution specification, to represent their success in solving the problem. In order to develop a marking scheme three expert analysts independently developed a solution, against which subjects' solutions were compared. The marking scheme contained a list of components to be included in a specification, and focused on semantic features of subjects' solutions rather than on the syntax of the data flow diagramming representation. These components included the processes, data stores, systems inputs, outputs and functional and non-functional requirements of the system. Subjects received a score if a component was included in the resulting data flow diagram or if the subject verbally stated that the component was to be included in the system. The composite expert solution to the delivery scheduling problem is given in Appendix A.

Results

All subjects physically drew or verbalised a solution to the problem in the time available to them. Twelve subjects developed data flow diagrams, all of which were incomplete. Subjects' overall performance was poor, averaging only 11.4% of the expert score.

Frequencies of mental behaviours were counted for 5-minute segments. No marked trends over time were apparent, although Information acquisition, Assertions and Planning decreased with time and Reasoning increased in the later stages of the protocols. Inter-individual differences were apparent in total occurrences of each particular behaviour.

No correlation was found between experience scores and solution completeness (Spearman rank order coefficient). Correlations between totals of behaviour and individual's experience of solution completion were also non-significant apart from Reasoning with completeness ($p < 0.05$, Spearman Rs). As this suggested reasoning ability may be linked to analytic success, reasoning behaviour was investigated in more detail.

Hypothesis life histories (generation, development, testing of hypotheses) were analysed to develop network models of reasoning behaviour. The total number of arcs between each pair of categorised reasoning utterances (e.g. generate hypothesis → modify hypothesis) were counted for each subject, and the total number of arcs for all subjects is represented graphically in fig. 1. Reasoning strategies could be discerned for 11 subjects by reference to these graphs and examination of the nature of hypotheses in the transcripts. All subjects used the weak Generate and Test method. From the protocol transcripts it was noticeable that subjects with low completeness scores only evaluated hypotheses with general tests which resulted in weak conclusions. Subjects with high completeness scores generated domain scenarios to evaluate hypotheses,

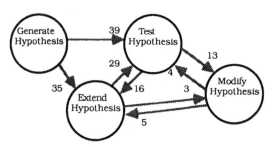

Fig. 1. Network model of hypothesis life histories for all subjects: figures on the arcs represent the total number of transitions which occurred.

suggesting effective testing of hypotheses may be a determinant of good analytic performance.

Planning behaviour was investigated in more detail. The majority of plans (97%) were short term describing the next sub-problem task and were executed within 5 minutes. When questioned about overall plans and strategies in the retrospective protocols all subjects had difficulty verbalising coherent plans, suggesting either the absence of long-term plans or the unconscious nature of such plans. Examination of hypotheses suggested analytic behaviour was triggered by reading the contents of the problem narrative.

Conceptual modelling (drawing the physical data flow diagram) was analysed by investigating Reasoning with hypotheses sharing closely related subject matter (i.e. linked to the development of a conceptual model by a single thematic strand). This behaviour was termed model-based reasoning, and was exhibited by six subjects. These subjects constructed more complete solutions than subjects who were judged not to be model-based reasoners, hence construction of conceptual models appeared to be related to improved analytic performance.

Analytic strategies of novice software engineers were investigated by examining their Planning and Reasoning behaviours in combination of the types of knowledge employed. Concurrent and retrospective protocols were examined for particular phases in analytic strategies. All subjects attempted to scope the problem but eight were ineffective in doing so. Problem scoping included identification of the system boundaries and recognition of key system requirements. Poor scopers became embroiled in the system details before determining the problem space. Five poor scopers may have been misled by SSA method heuristics, and immediately attempted to identify software engineering features (e.g. data stores). Good scopers had high completeness scores, suggesting that effective problem scoping may be another determinant of analytic success.

Only three subjects overtly employed method heuristics throughout the protocol. Most subjects proceeded to 'divide and conquer', analysing one area in depth before advancing to the next. Three subjects also disobeyed SSA dictates, by mixing analysis and physical design details in their solution. One subject developed the most complete conceptual model with several sort routines at its core, and there was no evidence that subjects were impeded by the inclusion of implementation features.

Subject(s)	Good problem scoping	Conceptual modelling	Critical testing
E	√	√	
G		√	
I	√	√	√
J	√	√	√
K	√	√	√
N		√	√
P	√		√
D, H, M, O R & S			

Fig. 2. Summary of possible reasons for poor and good analytic performance.

Conclusions

This study suggested good analytic performance was determined by effective hypothesis testing, model-based reasoning and good problem scoping during early phases of analysis. However, contrary to previous studies (e.g. Vitalari and Dickson 1983), results were individually variable (see fig. 2). Some weaker subjects exhibited behaviour to suggest that they relied on method knowledge to structure the analytic task, while better subjects were able to exploit domain knowledge to develop scenarios, structure the problem space and reason more successfully. Our proposed applied mental process model of systems analysis is described graphically in fig. 3. Software reuse is one means of providing novice software engineers with domain knowledge, hence we further investigated software reuse by novice and expert software engineers.

Study 2 – An initial study of software reuse by novice software engineers

Reuse of domain knowledge can occur between analogous problems, so an experiment was designed in which software engineers were required to reason analogously during software reuse. Specification-level reuse was investigated, to test whether software engineers could exploit a repository of potentially reusable specifications available in a CASE tool, and to investigate reuse as a means of assisting software engineers during systems analysis. The study also investigated the mental processes and knowledge structures involved in analogous reasoning during specification reuse.

The use of abstraction in the construction of conceptual models, represented as data flow diagrams or entity life histories, to describe application domains at a high level is widely advocated by software engineers. Furthermore, the use of abstract templates and generic objects has been proposed as a method of delivering reusability. In the light of this we investigated the reuse of specifications presented in concrete and abstract forms, as concrete presentation of analogy is often advocated in HCI (for example direct manipulation interfaces) whereas abstraction is valued for problem

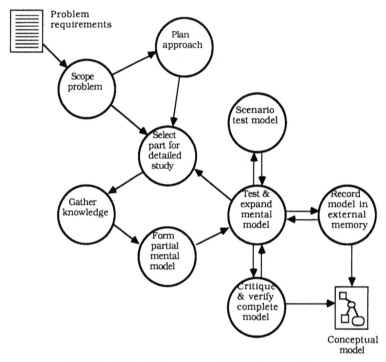

Fig. 3. Applied mental process model of systems analysis.

solving in software engineering. A detailed report of this study is given in Sutcliffe and Maiden (1990).

Method

The 30 subjects (23 M, 7 F) were full-time MSc students in Business Systems Analysis with little or no commercial analytic experience. They had knowledge of several structured analysis and Jackson (JSD) techniques (Jackson 1983). All but 6 of the subjects had previous systems development experience. Subjects were asked to develop a JSD process structure diagram for a scheduling function allocating videos to hotels in a video hiring company. An expert solution to this problem is given in Appendix B. The problem built on domain knowledge already acquired by the subjects from a case study.

Two analogies with the video hiring problem were purposely constructed. The concrete analogy was with a production planning system allocating manufacturing machines to production jobs. The abstract analogy described a scheduling function allocating resources to tasks which had to be fulfilled. The main analogical concept was the functional requirement to allocate a resource within certain constraints. This was manifest as scheduling a resource (video copies) within constraints (e.g. time, hotel preference, etc.). The reusable specifications were represented using JSD notation.

A between-subjects, two-condition experiment was conducted with:

- a Control group, where subjects were given the problem narrative alone,
- an Abstract Analogy group (group AA), where subjects were provided with the problem narrative and the abstract template,
- a Concrete Analogy group (group CA), where subjects were given the problem narrative and a JSD specification of the real analogous production planning application.

Each group of 10 subjects was balanced with respect to subjects' experience. Both reusable specifications were similar in size and complexity to the solution described in Appendix B.

In concurrent protocols lasting 1 hour Group CA and AA subjects were requested to verbalise reasoning and recognition of analogous similarities and the transfer of knowledge from the analogy. A video camera captured all written work, and the verbal protocols were tape recorded. Retrospective protocols (similar to that in study 1) and a written questionnaire captured problem solving and reuse strategies. Retrospective questioning probed subjects' general problem solving strategies by asking them how they achieved their solutions. Questions were directed at subjects' understanding of the reusable specification, the problem domain and the analogy. Subjects were also asked whether they recognised analogical links between the reusable and problem domains.

Subjects' success was measured by marking each solution for completeness as well as validity. Each solution was allocated a completeness score by comparing it to the expert solution in Appendix B. The marking scheme focused on semantic features of subject's solutions rather than on the syntax of the JSD representation. Subjects received a score if a component in their solution was included in the expert solution. The validity of solutions was measured by the quantity and severity of errors. Errors were counted and ranked on a 1–8 scale according to their severity, by examining the extent to which the specification was incorrect in terms of domain knowledge and JSD syntax. Solutions were independently cross-marked by two experts, who agreed on scoring in 91% of all cases for completeness scores.

Results

All subjects attempted a solution to the video hiring problem. Reuse of analogous specifications appeared to improve the completeness of solutions, and reusable material presented in an abstract form enhanced performance more than presentation of concrete specifications (completeness was measured as a percentage of the expert's solution – the Control group scored on average 24.4%, those given the concrete analogy scored 32.8%, whilst those given the abstract analogy scored 41.1%). This effect was significant for the abstract analogy (t-test using the approximating Z distribution for non-normal populations; $Z = 2.23$, $p = 0.05$). This may be because abstract templates were more easily recognised as analogous (all subjects recognised the abstract analogy, 80% recognised the concrete analogy). Although specification reuse improved completeness, it resulted in similar error rates hence reuse does not appear to help the creation of more accurate specifications.

Subjects were divided into successful and unsuccessful reusers, based on a minimum completeness score for their solutions and the explicit transfer of key features of the reusable specification. Eight group AA and five group CA subjects successfully exploited the analogy. Completeness and error scores for these subjects indicated that those who reused the concrete analogy produced more complete and valid solutions, although differences between groups were non-significant. These results suggested that reuse from concrete analogies may have been more effective once the analogy had been recognised.

Analysis of errors made during knowledge transfer revealed importance differences in the mistakes made by group AA and group CA subjects. The number and naming of key components by the AA subjects were closely related to the analogous template, suggesting they may have copied the material rather than reasoning about it. Specification copying in the sense of lexical tailoring of specification components, without reasoning, accounted for many errors. Errors in eleven of the 13 successful subjects' solutions suggested a general failure to understand the analogous specification. Retrospectively subjects admitted to transferring key concepts without understanding them, or omitting to transfer critical components which they did not understand.

The failure to understand the analogy led to mappings based on surface similarities between the problem and the reusable specification by 11 of the 20 group AA and CA subjects. Subjects appeared unable to construct mappings where no surface similarities existed.

Conclusions

This study suggested that novice software engineers were mentally lazy during software reuse, indicating that software engineers may adopt reuse as an easy cognitive strategy when given the opportunity. Abstract analogies appear to be easier to recognise and assimilate, however reuse of concrete analogies appeared to be more effective, possibly because the subject is forced to reason more about the analogy rather than syntactically copy the reusable specification. Novice software engineers are unlikely to have many domain analogies to draw on when constructing new mental models of a problem domain. It is therefore not surprising that when presented with an analogical prompt they take it as a potential ready-made solution.

Results suggested the need to investigate the analogous reasoning behaviour of novice software engineers in more detail. A comparable study of the analytic strategies and analogous reasoning behaviour of expert software engineers was also undertaken, to identify shortcomings in novice's analytic behaviour.

Study 3 – A study of reuse by expert and novice software engineers

Protocol analysis was used to investigate the cognitive processes during reuse of concrete analogous specifications. Expert and novice software engineers were requested to reuse a specification to develop a specification for an analogous problem domain. Detailed reports of this study are given in Maiden and Sutcliffe (in prep.-a, -b)

Method

Protocol analysis was used to investigate analytic and analogous reasoning behaviour of 10 expert and 5 novice software engineers. The 5 novices (3 M, 2 F) were MSc students in Business Systems Analysis, with a maximum of 6 months experience of SSA techniques, while the experts (8 M, 2 F) had up to 20 years commercial analytic experience, and were drawn from four commercial and academic backgrounds. Five analysts (4 M, 1 F) worked in local government and had knowledge of structured analysis techniques, although only one had employed these techniques on a daily basis. Four analysts (3 M, 1 F) employed by 2 consultancy firms had regularly used structured analysis techniques. The 10th expert (1 F) lectured in computing at an academic institution, and had experience in applying and teaching structured analysis techniques.

Subjects were asked to use Structured Systems Analysis (SSA) techniques (De Marco 1978) to develop a specification for an air traffic control (ATC) system. They were assisted by the provision of a specification describing a flexible manufacturing system (FMS) which was analogous to the ATC problem. All subjects had background method knowledge, and were provided with the domain knowledge necessary to develop a specification.

All subjects were given a narrative describing the air traffic control system, and a specification describing the analogous FMS. The 500-word problem narrative described the domain and functionality of the control of air traffic and requirements for the computerised system. The analogous FMS specification was represented using data flow diagramming (DFD) notation (see Appendix C) supplemented by short narratives describing the objectives and main processes of the system. Subjects had access to both documents at all times during the protocols, but were not allowed access to any other material.

Subjects were requested to think aloud and their verbal protocols were captured by a video camera which also recorded drawing and reading behaviour. During the protocols subjects were advised to take their time, and not be afraid of verbalising too much. Instructions for subjects were read by the experimenter. Each subject was strongly recommended to reuse the analogous prompt to develop two new data flow diagrams, because the problem was otherwise to difficult to complete in the time allowed.

Subjects were given 75 minutes to develop two data flow diagrams, as a pilot study indicated this was sufficient time to complete the task. All subjects were informed of this time limit before beginning the task, and were expected to complete a solution by the end of it. While each subject performed the task a concurrent protocol was recorded. Upon completion of the task verbal retrospective protocols and a written questionnaire elicited further details of subjects' analytic strategies and mental and non-mental behaviour. First a 10-minute written questionnaire elicited subjects' understanding of the analogy then the experimenter verbally questioned the subject for 15 minutes to elicit further details of subjects' analytic and reasoning strategies and to investigate specific hypotheses and errors. Retrospective questioning was controlled by a checklist of different behaviours which were expected to occur during the task (e.g. 'Why did you stop analysing the problem and start to develop your solution ?'). Care was taken not to prejudice the retrospective protocol, so the experimenter only asked open-ended questions, following Ericsson and Simon's practice. Finally all subjects completed a questionnaire identifying details of all previous analysis and programming experience.

The analogy between the ATC and FMS domains was carefully constructed to allow considerable reuse of the FMS specification, although several similar features of the reusable specification were 'red herrings' included to identify mental laziness during analogous comprehension and transfer. The main concept was the functional requirement to keep objects apart and to ensure that these objects follow a predetermined plan to their destination. This was manifest as aircraft protected by an air space which no other aircraft is permitted to enter and guided by a flight plan to its destination. Similarly products were protected by track sections which are only permitted to contain one product at a time and controlled by a production plan which directs each product.

Protocol transcripts were analysed twice: (i) by categorising the mental behaviours represented in speech segments, usually sentences and incomplete utterances; (ii) identification of analytic strategies using a taxonomy based on criteria of mental and non-mental activity. Non-mental behaviour represented physical activites exhibited by subjects, including reading, drawing and notetaking. Protocol utterances were categorised using definitions of mental and non-mental behaviour similar to those used in the first study. Mental behaviours were:

- *Assertions*: see definition in study 1;
- *Planning*: see definition in study 1;
- *Reasoning*: verbalisation of the creation, development and testing of hypotheses about the problem, its proposed solution or the source domain (e.g. 'the aircraft risk colliding, hence the warning process must be automatic, and it must inform the air traffic controller with warning messages displayed on the radar screen'). Each reasoning was further categorised to identify subjects' topic focus: (i) reasoning about the target (ATC) domain, (ii) reasoning about the source (FMS) domain, (iii) reasoning about analogous concepts between the source and target domains, and (iv) reasoning about general concepts which do not describe the target or source domains, or the analogous links between them.

Non-mental behaviours were:
- *Information acquisition*: searching for and retrieval of data in the problem text or the reusable specification;
- *Conceptual modelling*: physical construction of the system specification, recorded as a data flow diagram.

Non-mental behaviour was categorised as occurring concurrently with mental behaviour. Analytic strategies were based on mental and non-mental behaviour, and the purpose of the subjects' activity. They were classified using eight strategies:

- *Gather information*: read the target document or the reusable specification;
- *Summarise data*: summarise the contents of the target document or the reusable specification:

- *Reuse*: reuse the FMS specification to develop a conceptual model representing the solution to the ATC problem;
- *Construct*: develop a conceptual model representing a solution without reusing the FMS specification;
- *Revise*: redraw the solution;
- *Evaluate against the target*: test the subjects' solution against the target requirements in the problem document;
- *Evaluate against the analogy*: test the subjects' solution against the reusable specification;
- *Summarise solution*: test the subject's solution without accessing the problem document or the reusable specification.

Protocol categorisation was validated through cross-marking by two independent observers with experience of protocol analysis. Each observer allocated a behavioural category to each utterance in 3 randomly-selected protocols. Inter-observer agreement was 83% of all categorised protocol utterances, and differences between observer categorisation were reconcilable. Analytic strategies in three different protocols were also independently categorised, and the observers reconciled differences between categorisation of strategies to develop a common definition of analytic strategies in those and the remaining protocols.

Completeness and error scores were allocated to each subjects' solution to evaluate their success or otherwise in solving the problem. In order to construct a marking scheme, an expert solution was developed by two expert analysts. The marking scheme contained a list of components to be included in a specification, and focused on semantic features of subjects' solutions rather than on the syntax of the data flow diagramming notation. These components included the processes, system inputs and outputs, external entities and data store accesses in the expert solution. Subjects received a score if a component was included in the resulting data flow diagram, and each subjects' completeness score represented the number of required components included in their solution. The larger of the two reusable DFDs describing the FMS domain is given in Appendix C, and is similar in size and complexity to the expert solution to the ATC problem.

Subjects' solutions were also analysed to determine their validity. Specifications were examined for their inclusion of 5 types of syntactic error and 7 types of semantic error. Subjects received a score for each type of error included in their solution, and their score represented the total number of different error types made by each subject.

Results

All subjects attempted a solution to the ATC problem. Experts developed marginally more complete and valid solutions (see table 1) although there were notable differences between the success of individual experts and novices. Two novices (N3 and N5) produced a more complete solution than other novices, while one expert (E11) performed more poorly than all other subjects. E11 mistrusted the analogy and used method heuristics to limit the scope of the problem space and develop an incomplete solution.

Table 1
Completeness and error scores for all subjects.

Experts			Novices		
Subject	Completeness score (% age of expert score)	Number of error types made	Subject	Completeness score (% age of expert score)	Number of error types made
E1	59.1	2	N1	61.4	2
E3	79.5	1	N2	50	3
E4	75	1	N3	72.7	3
E5	70.5	1	N4	54.5	2
E7	63.5	1	N5	72.7	0
E8	70.5	1			
E9	68.2	2			
E10	61.4	2			
E11	38.6	3			
E12	61.4	1			
Average	64.8	1.5	Average	62.7	1.8
Range	38.6–79.5	1–3	Range	50–72.7	0–3

Analytic strategies were examined to identify patterns of analytic behaviour. The occurrence of analytic strategies were counted within each 5 minute time period (see fig. 4) The trends in frequencies of strategies suggested that a period of problem scoping occurred before the solution was built and tested. Frequencies of mental behaviours were also counted for 5 minute segments. Initial bouts of Information gathering during problem scoping led to the peak in the frequency of assertions at the beginning of the protocols. Results indicated that all subject's tended to scope and gather information about the problem during the first part of the protocol, and test their solutions during the latter stages. Subject's behaviour over time also revealed expert-novice differences. Experts began conceptual modelling before novices, suggesting that experts may be able to scope the problem and develop a solution more quickly. This appears to be caused by expert's tendency to construct a solution during the first 25 minutes of the protocol. Closer examination of protocol transcripts revealed that some experts attempted to directly develop the first, simple required conceptual model, however, analysis of their completeness and error scores revealed that they were less successful than reusers. Examination of the use of testing strategies revealed that novices tended to evaluate their whole solution against the problem requirements at the end of the protocols while experts appeared to test each model after it had been developed. Increases in expert bouts of Evaluate against the target strategy led to an increase in the frequency of assertions between 50 and 65 minutes.

Experts exhibited more Reasoning behaviour than novices, suggesting as in the first study that reasoning ability may be linked to expert analytic behaviour. A sequential analysis (see study 1) of detailed reasoning behaviour was carried out by examining hypothesis life histories. All subjects exhibited a Generate and Test cycle (see fig. 5), however experts exhibited more extensive development and testing of hypotheses and

components\
uetterances:
average per
subject

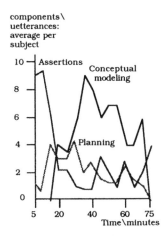

Number of utterances verbalised
and components added during a
5-minute period for novices:

components\
uetterances:
average per
subject

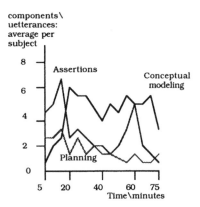

Number of utterances verbalised
and components added during a
5-minute period by experts:

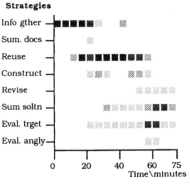

Number of novices using a strategy
within 5-minutes periods:
■ = 5 subjects ▓ = 1 subject.

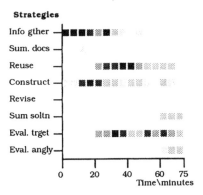

Number of experts using a strategy
within 5-minutes periods:
■ = 10 subject ▓ = 1 subject.

Fig. 4. Use of analytic strategies and verbalisation of mental behaviour over time.

were able to test more effectively (i.e. further extended hypotheses as a result of testing). Examination of protocol transcripts suggested that experts may be more able to reason about a problem, to create test cases and scenarios which can be constructively evaluate hypotheses.

Analytic strategies were examined in more detail to identify differences in expert–novice behaviour. The length of time of each strategic bout and and the reasoning

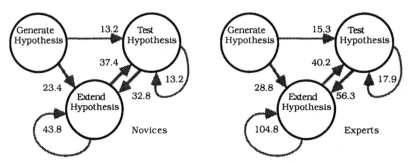

Fig. 5. Hypothesis life histories representing average number of transitions which occurred for experts and novices – only major reasoning categories are shown.

behaviour verbalised during each of these bouts was quantified (see table 2). Novices spent more time reusing the analogous specification and summarising their own solutions, while experts appeared to be stronger information acquirers and evaluated their solutions against both the reusable specification and the requirements document. Closer examination of testing strategies revealed that experts were more able to add to the completeness of their solutions during evaluation of solutions against the problem requirements. Improvements to solutions were additional requirements which had been omitted during building of the solution, and each expert increased the completeness of their solution by 7.6% while novices only managed an improvement of 0.89%. Experts also demonstrated greater improvement in their solutions during solution evaluation against the analogy. This suggested that experts were more effective testers and were more able to identify omissions and errors in their solutions.

The analogous specification was reused by all but one subject to develop a major part of each solution, so reuse behaviour was examined more closely. Investigations of subject's reasoning behaviour and analytic tactics during reuse revealed important

Table 2
Average lengths of time and reasoning utterances during reasoning strategies spent by novices and experts.

Analytic strategy	Av. times (min)		Analytic strategy	Av. reasoning utt.	
	Novices	Experts		Novices	Experts
Information gathering	19.6	21	Information gatherings	37.8	86.4
Summarise problem	0.4	0.3	Summarise problem	2.4	5.8
Reuse specification	24	14.6	Reuse specification	119	111.3
Construct solution	5.4	10.2	Construct solution	16.2	51.3
Revise solution	3.2	0.8	Revise solution	9	4.8
Summarise solution	8.25	3.3	Summarise solution	33.2	23.7
Evaluate vs. analogy	0.6	3.4	Evaluate vs. analogy	3.4	24.3
Evaluate vs. requirements	9.8	11	Evaluate vs. requirements	28.2	51.1

Table 3
Approaches used by subjects to read about and understand the reusable specification.

Information gathering tactics to understand the reusable specification	Number of experts (as % age)	Number of novices (as % age)
Read the DFDs only	10	40
Read the narrative and the DFDs	30	40
Read the narrative only	60	20

expert–novice differences. Eighty percent of the novices verbalised that they copied the analogous specification and directly substituted words to develop a solution. Novices who developed more complete solutions did so by copying the reusable specification in great detail and transferring as many candidate reusable components as possible, while subjects who used alternative strategies to complete their solutions were less successful. Alternatively 80% of experts stated that they deliberately avoided a copying strategy, and voiced concerns about reusing a specification that they did not understand. Examination of concurrent protocols revealed that all experts exhibited strategies and mental behaviour consistent with attempts to understand the reusable specification. This indicated that novices appeared to be mentally lazy while experts attempted to understand the analogy before reusing it.

Strategies to scope the problem were investigated in more detail by examining the analytic tactics used to understand the reusable specification. Information gathering from the reusable specification focused either on reading the DFDs or the supporting narrative. Results suggested that the experts favoured the problem narrative rather than the reusable DFDs to assimilate and understand the analogy while the novices concentrated on the DFDs (see table 3). This may have been linked to the type of knowledge contained in the narrative. The reusable specification contained solution knowledge representing the computer solution to the FMS problem, while the narrative described the underlying FMS problem domain, hence experts may have considered problem knowledge more important than solution knowledge in understanding the analogy.

Reasoning about the problem domain was investigated in more detail. Reasoning utterances were categorised as reasoning about (i) the subject's computer solution, (ii) the goals which the subject's computer solution must achieve, and (iii) the problem domains in which the computer system must operate. Experts reasoned more about the problem domain (average $30.2 \setminus 67.7$ utterances) and significantly more about the computer system's goals (average $4.5 \setminus 13.5$, t-test: $p = 0.037$). This supported other findings and indicates that experts may have developed more complete models of the problem domains which underlie the target and analogous specifications.

Subjects' understanding of the analogy was examined more closely during retrospective protocols. Subject's final understanding of the analogy was investigated by requesting target mappings for each object in the source domain. Surprisingly novices recognised more correct analogous mappings (average 53.7%/41.8%) however experts made many fewer erroneous mappings (average 30%/11.8%) hence experts appeared to

Table 4
Retrospective understanding of analogous mappings, prompted from candidate source domain mappings – italics identifies important analogous mappings perceived by subjects.

Analogous mapping		Experts		Novices	
Source domain	Target domain	Corr-ectly map-ped	Incor-recty map-ped	Corr-ectly map-ped	Incor-recty map-ped
Production Controller	*Air Traffic Controller*	9	0	5	0
Infra-red Sensors	*Radar*	9	0	5	0
Production Plan	*Flight Plan*	8	0	5	0
Production Track	*Air Corridor*	7	2	2	3
Product	*Aircraft*	6	2	4	1
Two products in same track section	*Two aircraft in same Air Space*	6	0	5	0
Manufacture of a Product	Flight	3	2	2	1
Product Type	Aircraft Type	3	3	1	2
Production Floor Layout	Airways	2	4	3	1
Misdirected Product	Aircraft Off Course	2	0	3	0
Production Track Section	Air Space	1	5	1	2
Job	Flight Step	1	2	0	4
Production Operator	Pilot	1	3	3	2
Machine	Air Space	1	1	0	2
Delayed Product Manufacture	Delayed Flight	0	2	3	0
Lost Product Manufacture	Missing Flight	0	0	0	1

have a better understanding of limited subset of the analogy. Subjects tended to recognise certain mappings more than others, and all subjects most often correctly mapped aircraft/products, flight plans/production plans, air traffic controller/production controller, radar/infrared sensors and two aircraft/products in the same space (see table 4). Experts also tended to correctly map air corridor/production track, while a greater precentage of novices correctly mapped 'airways' to 'production floor layout. This suggested that neither set of subjects had a very good understanding of the analogical mappings between the physical structure of both domains. Retrospective questioning also investigated the analogous mappings which subjects perceived as critical to the analogy. Results supported previous findings however novices exhibited a weaker understanding of the analogy than the experts, although experts understanding was only based on half a dozen analogical mappings. Subjects were also requested to generalise these critical analogous mappings, and experts were more able to generalise these key analogous mappings in abstract terms when requested (average 5 generalised mappings for experts, one for novices). Results suggest that subjects developed analogous links between mental models at different levels of abstraction; novices developed mappings between computer solutions (e.g. files processes) while experts recognised mappings between the underlying problem domains.

Expert's understanding of the analogy was examined in more detail by re-examining concurrent protocols to elicit reasoning about the critical analogous mappings, and the

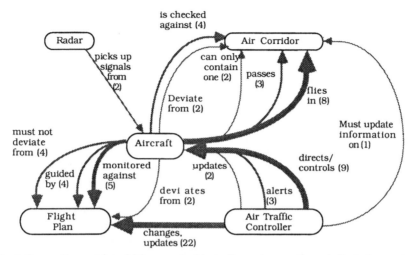

Fig. 6. Composite mental model for expert subjects: figures in parentheses indicate the number of times that a relation was verbalised by any subjects.

structure of the most commonly-mapped objects. Retrospective questioning elicited critical analogous mappings between objects, then reasoning behaviour during concurrent protocols was examined to identify major causal, functional and structural relations between mapped pairs of objects, similar to Gentner's structure-mapping theory (Gentner 1983). The model is represented as an informal semantic network in fig. 6. The analogous mapping between aircraft and product was central to expert's understanding of the analogy, and supports the model developed by experts when they were retrospectively requested to develop an entity-relation model of the domains. Relations linking objects were primarily functional hence experts understood the analogy in terms of the functionality of the domains rather than the physical structure of the problems. The most commonly-used relation was 'Air Traffic Controller updates/changes the Flight Plan', while only two structural relations were identified ('Aircraft flies in Air Corridor' and 'Aircraft passes along Air Corridor'). This poor understanding of the physical structure of the analogous domains supports retrospective findings and is likely to have led to incorrect analogous mapping identified during reuse.

Study 1 suggested the importance of individual differences in analytic behaviour. Expert and novices' analytic strategies were examined for common patterns of behaviour however none could be identified. All but one subject relied on reuse to develop a majority of each solution however subjects adopted different during during reuse and evaluation of their solutions. Results from study 3 supported those from study 1 and indicated that analytic behaviour was individually variable.

Conclusions

As in the first study results suggested that experts were better scopers and testers. However differences in the solutions developed by expert and novices were small, suggesting that specification reuse may have improved the analytic performance of novice software engineers to a near expert level. This conclusion is supported by the poor performance of novices in the first study, and the potential benefits of software reuse on the analytic process identified in the second study. Closer examination of analytic and reuse strategies revealed important differences in how subjects developed their solutions. Novices were mentally lazy and copied much of the specification while experts attempted to understand the analogy before reusing the specification. Experts appeared to concentrate and reason more on the underlying analogous problem domain, and to test their solution against both the target requirements and the analogy. However experts only developed a partial understanding of the analogy based on 6 analogous mappings. Novices also develop a partial but weaker understanding of the analogy, based on mappings between solution components rather than problem features. Experts appeared to concentrate and reason more on the underlying analogous problem domain, and to test their solution against both the target requirements and the analogy.

General discussion

The three studies reported in this paper can be discussed in terms of an applied mental model of analytic behaviour and use of analogical reasoning during software reuse.

The results from the first and third studies suggest that systems analysis is complex and individually variable. Determinants of good novice analytic performance include proper problem scoping, strong solution testing and development of well-formed mental models. The investigation of expert analytic behaviour in the third study suggested that reasoning about the problem domain underlying a computer system may be another determinant of good analytic performance. Results from the first study suggest that procedural method knowledge failed to support the analytic process, and novice software engineers who followed method dictates too closely may have been hindered from good problem scoping.

Results from the experimental studies suggest that reuse may have enhanced the performance of software engineers, although novices tended to copy rather than reason analogously with the reusable specifications. Copy-cat strategies are unlikely to support successful reuse of analogies since this involves more customisation of the specification, hence tutorial support (e.g. Polson and Richardson 1988) may be necessary to encourage and guide analogous reasoning. The third study suggested that even expert analysts were unable to fully understand the analogy hence the size and complexity of many analogies between software engineering problems may be an impediment to successful reuse. A realistic reuse scenario may be a CASE tool incorporated with an intelligent advisor (e.g. Cummings and Self 1989) which can 'reason' along side and prompt the software engineer. We are developing didactic strategies and tactics from detailed analysis of expert reuse strategies, and issue-based diagnosis may be based on software engineers bugs and misconceptions identified in the third study.

Results from the third study support several existing theories of analogy. Expert software engineers reasoned about the analogy in terms of functional rather than structural relations which described the domain, supporting Gentner' s structure-mapping theory (1983). Both experts and novices also generalised during analogous reasoning, supporting Greiner's (1988) model which purports that analogous transfer only occurs between concepts which share a common abstraction. Gick and Holyoak (1980) propose analogous reasoning between two problems is only possible if they have common goals. Many subjects recognised the analogy through similar system goals, hence similar goals appeared to support analogous reasoning. Existing theories of analogy are supported by simplistic examples. The scale of software engineering problems employed in these investigations suggests analogy may be more complex than implied by existing theories. Analogy may have goal, structural and abstraction components, each of which is a critical determinant of an analogy.

Appendix A: Expert solution for study 1

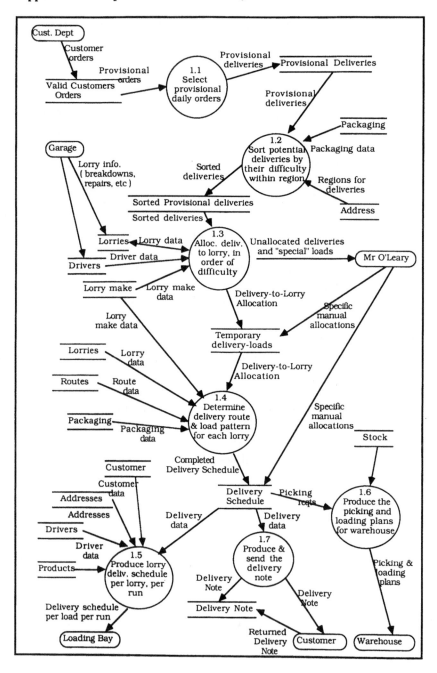

Appendix B: Expert solution for study 2

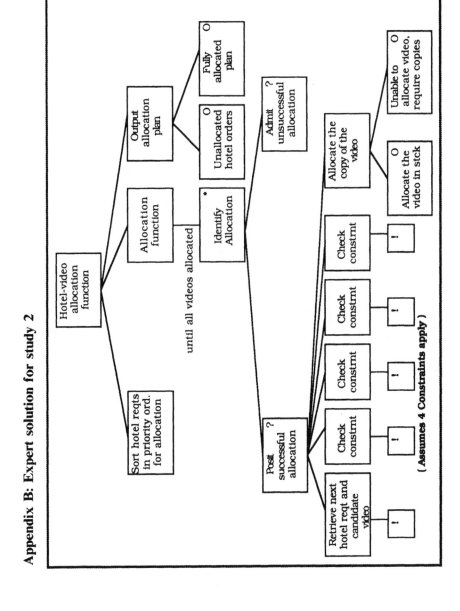

Appendix C: Larger reusable DFD for the FMS domain in study 3

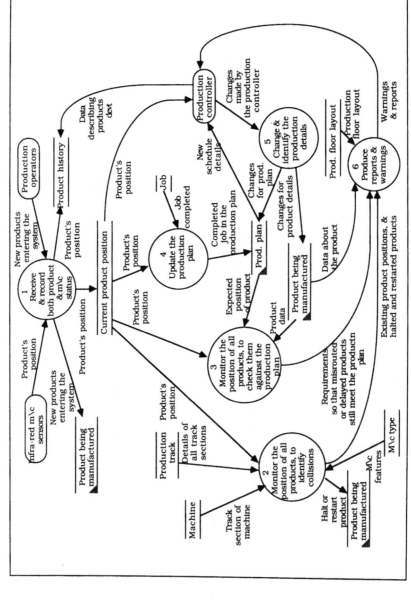

References

Adelson, B., 1984. When novices surpass experts: The difficulty of a task may increase with expertise. Journal of Experimental Psychology: Learning, Memory and Cognition 10, 483–395.

Anderson, J.R., 1983. The architecture of cognition. Cambridge, MA: Havard University Press.

Cummings, G. and J. Self, 1989. 'Learner modelling in collaborative intelligent educational systems'. In: P. Goodyear (ed.), Teaching knowledge and intelligent tutoring. Norwood, NJ: Ablex.

De Marco, T., 1978. Structured analysis and systems specification. Englewood Cliffs, NJ: Prentice-Hall.

Ericsson, K. and H.A. Simon, 1984. Protocol analysis. Cambridge, MA: MIT Press.

Gentner, D., 1983. Structure-mapping: A theoretical framework for analogy. Cognitive Science 7, 155–170.

Gick, M.L. and K.J. Holyoak, 1980. *Title of article?* Cognitive Psychology 12, 306–355.

Greiner, R., 1988. Learning by understanding analogies. Artificial Intelligence 35, 81–125.

Guindon, R. and B. Curtis, 1988. 'Control of cognitive processes during software design: What tools are needed?' In: E. Soloway, D. Frye and S.B. Sheppard (eds.), Proceedings CHI '88: Human factors in computer systems. ACM Press. pp. 263–269.

Jackson, M.J., 1983. Systems development. Englewood Cliffs, NJ: Prentice-Hall International.

Koulek, R.J., G. Salvendy, H.E. Dunsmore and W.K. Lebold, 1989. Cognitive issues in the process of software development: Review and reappraisal. International Journal of Man–Machine Studies 30, 171–191.

Maiden, N.A.M. and A.G. Sutcliffe, in prep.-a. The abuse of reuse: Why software reuse must be taken into care.

Maiden, N.A.M. and A.G. Sutcliffe, in prep.-b. Cognitive models of expert software reusers.

Newell, A. and H.A. Simon, 1972. Human problem solving. Englewood Cliffs, NJ: Prentice-Hall.

Polson, M.C. and J.J. Richardson, 1988. Foundations of intelligent tutoring systems. Hillsdale, NJ: Erlbaum.

Sutcliff, A.G. and N.A.M. Maiden, 1990. 'Software reusability: Delivering productivity gains or short cuts'. In: Proceedings of INTERACT '90, August 27–31, Cambridge, UK. pp. 895–901.

Sutcliffe, A.G. and N.A.M. Maiden, in press. Analysing the analyst: Cognitive models of software engineering.

Vitalari, N.P. and G.W. Dickson, 1983. Problem solving for effective systems analysis: An experimental exploration. Communications of the ACM 26, 948–956.

Database interrogation

Acta Psychologica 78 (1991) 201–225
North-Holland

An experimental study of the interpretation of logical operators in database querying

Peter J.M.D. Essens, Carol A. McCann * and Mark A. Hartevelt

TNO Institute for Perception, Soesterberg, The Netherlands

The use of logical operators in query languages is considered to be a major source of user problems in database querying. The present study investigated whether people untrained in logic could successfully interpret logical operators; and, how errors and latencies are related to the structure of the query. In an experiment, the logical complexity of an SQL-style query formulation was varied in using AND, OR, and NOT operators in either single or combined form. The latency and error data converged to show that subjects had increasing difficulty with queries constructed with a combination of different operators. The inclusion of parentheses had a strong positive effect on task performance. Verbal protocols were used to identify sources of errors in query processing. A model of query processing was formulated and predictions of latencies and errors on the basis of processing components were tested.

Database management systems (DBMSs) are now widely used for structuring and storing textual information in applications ranging from small business administration systems to large library cataloguing systems. The user of a DBMS retrieves information from the database by specifying the subset desired along dimensions that are recognized by the database. This involves two major steps. The first is the formulation of a mental representation of the information subset that is desired, based on the user's conception of the way in which the database is structured. In the second step, that representation is transformed into an explicit query using a database query language or environment. The subset defined by the query is then extracted by the DBMS and presented to the user who checks to see if it is the desired

* On scientific exchange from the Defence and Civil Institute of Environmental Medicine, North York, Ontario, Canada.

Requests for reprints should be sent to P.J.M.D. Essens, TNO Institute for Perception, Kampweg 5, 3769 DE Soesterberg, The Netherlands.

set. If it is not, the loop of query creation, submission, extraction and checking is repeated until the user is satisfied.

A variety of interfaces and querying languages have been developed to provide access to data sets (Vassiliou and Jarke 1984). Many are based on the structure of the language SQL (Chamberlin et al. 1976). Yet studies of such interfaces have demonstrated that people, especially non-programmers, have difficulty using them (Jarke and Vassiliou 1985). To date, the fundamental research investigating database querying has focussed mainly on the following areas and issues: the user's mental model of the information system; influence of the structure of the information set; the syntax of the query language; and the use of Boolean operators for subset specification, the topic of this study. The remainder of this introduction gives a brief description of the first three issues mentioned; this is followed by a consideration of research done to date on the use of logical operators in querying.

Users are normally taught procedures for using an information retrieval system, for example, by formal training or from a manual. However, after this training, they may still have a poor mental model of the underlying capabilities, structure, and internal relationships in the information system. This lack of appropriate mental model prevents users from making inferences and predictions about the system's behaviour and inhibits further learning. In a study addressing this issue, Borgman (1986) demonstrated experimentally that subjects trained on simple retrieval tasks via a 'conceptual' model of an on-line library catalogue were better able to perform complex tasks that required extrapolation from the basic concepts than those subjects who were trained only on retrieval procedures.

The data contained within an information system are conceptually organized by the designers according to a 'data model'. There are three generic frameworks in which data is typically structured: the hierarchical, network and relational frameworks (corresponding to the different types of database management systems). The particular subdivision, grouping and linkage of a specific set of data within the selected generic framework forms the data model. Problems can arise if the structure of the data model is unknown to the user or is inappropriate for the task. Early work by Broadbent and Broadbent (1978) investigated the structuring of classes of objects and the allocation of descriptor terms to the classes. It showed that a person's retrieval of information is much better when the queries are based on the terms that he or she had assigned. Later studies (Lochovsky and Tsichritzis 1977; Brosey and Shneiderman 1978) compared the use of different generic data models on query writing.

Retrieval problems are sometimes due to factors at a lower level of human–computer communication, for example, the particular syntax or lexicon used in the query language itself. These issues have been identified in language usability studies carried out during the design of new query environments. Such studies have identified minor, but repeatedly-made errors like omitted punctuation, misspelled terms and the incorrect use of synonyms (Reisner 1977; Welty 1985; Thomas and Gould 1975).

In most database interfaces, the user specifies the subset of information that is desired by using Boolean connectives to form a logical combination of attributes that describe the set. Problems can arise when the user's concept of subset formation does not match the Boolean-based one demanded by the computer. The current work is focussed on this particular issue in information retrieval.

There is much incidental and anecdotal comment on the difficulty people have with logical (Boolean) operators like AND and OR in database querying (e.g., Thompson and Croft 1989; Cooper 1988). Marchionini (1989) notes that the Boolean operators provided for filtering the retrievals from an electronic database like an encyclopaedia are important tools for effective use of these information systems. However, in a study of the use of such databases by students, he found that they did not seem to take full advantage of search tools involving Boolean connectives. The AND operator was used as a connective in only one third of the electronic searches; the OR and NOT operators were never used. In the domain of library retrieval, Borgman (1986) found that 25% of subjects learning an SQL-like query language could not pass benchmark tests for system proficiency, although these tests were representative of the searches that were supported. The problem seemed to lie in the use of Boolean logic: more than one quarter of the subjects could not complete simple search tasks involving the use of one index and at most one Boolean operator.

One aspect of the difficulty stems from an incompatibility between natural English usage of the connectives 'and' and 'or' and their use in database retrieval. Ogden and Kaplan (1986) investigated this problem in detail, showing that the English word 'or' is most frequently used to indicate union, but that 'and' is often used ambiguously to indicate both union and intersection. Thus the statement 'Show the students in grades 10 and 11' implies the union (logical conjunction) operation (the set of students in grade 10 plus the set of students in grade 11), despite the use of the word 'and'. Since people often attempt to 'translate' the English-language statement of a problem into the query language in a phrase-by-phrase way, the different meanings of the connectives are not taken into account (Reisner et al. 1975). Ogden and Kaplan propose the incorporation of simple syntactic elements in the retrieval language to permit users to clarify ambiguous logical combinations.

The other difficulty in subset extraction has to do with the actual understanding and use of the logical operators in subset specification. This notion has not been as thoroughly investigated. Wason and Johnson-Laird (1972) found that subjects could more easily describe conjunctive concepts (involving AND) than disjunctive ones (involving OR) and that concepts which involved both conjunctive and disjunctive relations were the hardest to describe. Furthermore, subjects seemed to prefer positive descriptions of concepts rather than negative ones, even though the negative description was more efficient. The results suggest that people not formally trained in logic have difficulty with specific logic concepts.

The results of these studies of logical reasoning seem to be borne out in a more recent study by Greene et al. (1990) in which performance on subset specification using an SQL-type query language with explicit Boolean operators is compared with a tabular interface. In one condition, subjects were required to generate simple queries involving AND, OR, NEGATION, and AND + OR. In another condition, they had to choose the SQL query corresponding to a certain English statement. The average number of correct queries generated using SQL was only around 55%, except for the AND queries which were less error prone (73% correct). The time required to generate the query increased in the order AND, OR, NEGATION, AND + OR. Fewer errors were made in the choose condition, but the pattern of time required for selection was the same.

A study by Katzeff (1986) achieved higher success rates for generation of the same kinds of queries using a language essentially equivalent to the Boolean expression part of SQL. Subjects were first trained to interpret query specifications of the following logical forms:

P, NOT P, P AND Q, P OR Q, NOT (P AND Q), NOT (P OR Q)

Query interpretation required subjects to determine whether a given query would retrieve the information presented on an accompanying Venn diagram. They were then tested on query formulation for these same forms of queries and additional new forms:

P AND NOT Q, P OR NOT Q, NOT P AND Q, NOT P OR Q

With feedback and as much time as needed, subjects were able to generate queries in the first set with a success rate of 96%. Furthermore in almost half of the cases in the second set (45%) they were successful in identifying and using a single statement to express the logical restriction, although they had not been trained on them.

In a third investigation, Michard (1982) compared a Venn diagram aid for subset specification with the classical explicit Boolean operators as in SQL. The query problems involved three or four clauses, and various combinations of AND, OR and MINUS. The SQL specification gave rise to four times as many errors in set specification (38%) as the Venn specification method, including errors such as missing parentheses and criterion, and incorrect operators.

Although none of the latter three studies focussed specifically on the issue of logical operators in querying, they provide some evidence to confirm the less rigourously derived observations of Marchionini and Borgman, that, indeed, users are often not successful in correctly employing logical operators in SQL querying. However, the results are somewhat contradictory, with the Greene et al. experiment giving rather higher error rates for SQL query formulation than that by Katzeff. There are several important methodological factors that prevent direct comparison of the results, and, in addition, preclude the drawing of conclusions about precisely where in the querying process the problem with the logical operators actually lies.

All experiments involved the generation of SQL-style queries from test questions posed in natural language. However, the exact task of the subject varied: in the Michard study, subjects were required to create the whole query; in the others, only the part involving the logic of subset specification. Even with liberal error scoring methods, as used by Greene et al., it is possible that many of the errors in query formulation were due to factors additional to the understanding of logical operators per se, for example, the subjects' memory for details of syntax. It should be noted that the Greene et al. experiment circumvented this complication by having a separate task condition that required only the recognition of the correct query from a set.

All studies investigated the use of the operators AND and OR, although there were differences in the logical combinations used and in the overall 'complexity' (number of operators and structure) of the queries demanded. Two of the experiments tested the NOT operator, whereas the material used by Michard included the MINUS operation instead. The precise logical form of queries used in the experiment by Greene et al. was not described. The particular combination of logical operators has a large influence on the processing of the query. This factor was deliberately manipulated in the current study.

A further complication in query formulation concerns the influence of the natural language stimulus material on query formulation, especially given the potential confusion between natural language usage of logical connectives and their use in databases as elucidated by Ogden and Kaplan. In addition, there is the more general problem of determining whether the subject actually knows which information subset is requested by the stimulus question. The Greene et al. study handled this issue nicely by including a confirmatory test phase to check on subjects' understanding of the test question.

An additional difference in task across studies concerned feedback on response. Subjects were given feedback about the correctness of the query in the Katzeff study, and encouraged to try again on queries incorrectly specified. This may explain the better performance in this study compared to that by Greene et al. or by Michard.

Training for subjects was by the 'standard' SQL manual with examples and exercises in the Greene et al. experiment; by the use of written instructions and Venn diagram test examples in the Katzeff study; and by demonstration of examples in the Michard study. The form and extent of the training may well have a strong influence on subjects' comprehension of the meaning and use of logical operators. Special attention was paid to training and feedback in the current experiment.

In each previous study, subjects had no experience in computer use, although it is possible that they may have had training in logic, which would have influenced ability to do logical manipulations. The level of general education was higher in the Katzeff and Michard studies (university/college versus highschool).

In every case, the experimental part of the study was relatively short, involving the generation of only between 8 and 24 queries. Thus it is not possible to determine how performance using logical operators might change due to more extensive use of the language. Our study tested performance over a long series of trials.

In summary, the studies to date have not specifically addressed the cognitive interpretation of SQL logical operators under well-controlled conditions, in particular the relationship between errors and the specific operators and logical structure of a query. The current study, therefore, is aimed at providing more insight into what is the source of difficulty in the use of Boolean operators in subset specification in querying. More precisely, we wanted to see

(a) whether people untrained in logic could successfully interpret the logical operators (i.e., intersection, union, negation) as used to specify subsets in standard database querying;

(b) what kind of errors they made and how the errors related to the structure and complexity of the subset specification.

Method

There were several considerations that influenced the design of the experiment. First, as this was an experimental investigation, we wished to keep the subject's task as simple as possible. We wanted to avoid, for example, the complications introduced by use of 'and' and 'or' in natural language. Furthermore, we wanted to concentrate on the subject's conception of Boolean operators per se and to avoid the influence of complex processes like language transformation assumed in query writing (Reisner 1977). We therefore chose query interpretation rather than query writing as the focus of attention, the rationale being that the latter involves the former. In addition, the dataset attributes were limited to four in number to control for the learning of the data model. Second, since it's use is now so widespread, we decided to employ an SQL-style of presentation as the subset specification 'language'. Third, we choose 'naive' subjects – those without training in programming or logic. Furthermore, we limited the amount of training provided to subjects to the minimum necessary to get them going on the task, in the expectation that this strategy would better reveal the errors that they made.

The general task of the subject was to decide whether the SQL-style query presented on the screen would or would not result in the selection of the data base item that was described along with the query. Performance was measured in terms of latencies to evaluate each query and the errors made. Verbal protocol sessions gave qualitative data concerning the query evaluation process.

Subjects

Twelve students between the ages of 18 and 28 years participated in the experiment. Four of them were university students; the other eight were finishing or had just

```
┌─────────────────────────────────────────────────────────────────────┐
│         not artist = "Will Bazar" and year > 1970 or genre = 'pop'    │
│                                                                       │
│         TITLE          ARTIST      GENRE  YEAR  RATING                 │
│                                                                       │
│  YES  NO   Can you feel  Nenny Red   jazz    1969   3                  │
│   ┌──────┐                                                            │
│   │ Next │                                                            │
│   └──────┘                                                            │
└─────────────────────────────────────────────────────────────────────┘
```

Fig. 1. A sample stimulus showing the query (top line), the LP-instance (bottom line), response buttons, and a button to call the next stimulus.

finished high school. They were selected on the basis of absence of experience with computer programming or any other substantial knowledge of logic. The subjects were paid for their participation.

Stimuli

The stimulus set was constructed in two steps: first, a set of generic queries differing in the combination of logical operators was created; second, (arbitrary) information about different musical recordings ('LP's') was connected to these queries. The LP-instance consisted of five characteristics or attributes: title of the LP, name of the artist, genre of music, year of recording and rating of the LP. The query was an SQL-type subset specification with two or three conditions ('clauses') referring to LP-instance. The clauses, which in some cases were negated by the logical operator NOT, were connected by logical operators AND or OR. For each trial, a query and an instance of an LP were presented. The subjects had to decide whether the LP-instance did or did not match the query. The operator 'AND' means that both clauses must be true for the instance to match the query. The operator 'OR' means that one or both clauses should be true for a match. A more complex sample stimulus is shown in fig. 1. In the example the LP-instance does not match the query: although the artist is not Will Bazar, the year is not later than 1970; furthermore, the genre does not match. The title of the LP was never used in the query. A notation employed henceforth in the paper to describe queries uses A, B, and C to represent the clauses in the query; thus the sample query can be written NOT A AND B OR C.

The combinations of logical operators used in this experiment are shown in table 1. There were four main query categories consisting of two subcategories with combinations of logical operators. Each main query category was crossed with three levels of NOT (none, one or two). The position of the NOT(s) was varied across the A, B, or C clauses. The 'mixed' case appeared with and without parentheses. Parentheses were used only in combination with the second logical operator, i.e., NOT A OR (NOT B AND C). No default operator precedence rules were used, the rationale being that this would add an additional degree of complexity. Thus AND and OR were treated as equal in priority. Parentheses were used to indicate grouping of the clauses.

Two main sets of 34 stimuli representing the range of combinations of logical operators and having different query content were formed. The within-clause compara-

Table 1
Schematic representation of the 24 combinations of logical operators used in the experiment.

	Query categories			
	Single	Double	Mixed	Mixed( )
	Logical operator combinations			
	AND;OR	AND AND;OR OR	AND OR;OR AND	AND (OR);OR (AND)
0-NOT				
1-NOT				
2-NOT				

tors ($=$, $< >$, $<$, $>$) were evenly distributed over these queries. Unintended peculiarities of the LP-instance in the query part of the stimulus were controlled by varying the position and/or content of the LP-instance between the two subsets of stimuli. Queries in which the logical operators differed only in the position of the NOT were distributed over the two subsets in order to reduce the number of stimuli. The two subsets were replicated four times for a total of eight blocks. The content of the LP-instance part of the stimulus was varied between blocks so that the number of 'yes' and 'no' responses was evenly distributed.

A third set of 40 stimuli (for the verbal protocol sessions) representing most combinations of logical operators, and having the same constraints as the two main sets was also formed. The full set of logical combinations in this set is shown in table A.1 of the Appendix.

Procedure

The task of the subjects was to evaluate whether a certain LP described in the LP-instance would be selected by a given query. Latency and error data were collected in a so-called 'speed session', in which subjects were instructed to perform the task as quickly as possible. The speed session consisted of the eight blocks of 34 (randomly ordered) stimuli. Between the blocks was a short pause. Before and after the speed session there was a 'verbal protocol session' in which the subject was asked to think aloud during the evaluation of the queries. The two blocks of the verbal protocol sessions were completely identical and consisted of 40 (randomly ordered) stimuli each.

At the start of a session, subjects were given written instructions about the experiment and the task. The task was explained primarily by examples. No reference to truth tables or the explicit truth or falseness of the clauses in the examples was made. No specific reference to precedence of operators was made. Subjects were told that parentheses meant a grouping of the clauses. A series of sample queries were presented, for instance:

artist = 'Will Bazar' and year > 1970 or genre = 'pop'.

The explanation accompanying this query was 'The selected LP's have to be made by Will Bazar after 1970 or be pop LP's.' A list of six LP-instances was presented with the query and the result of every comparison of query to instance was given for the subject to study and confirm. The effect of the AND and OR were summarized in the following manner: 'When an AND is used, then the LP has to have both characteristics to be selected. When an OR is used, then the LP must have one (or all) characteristics.' To test whether subjects had learned the logical operators adequately, twelve test queries (six on paper and six on the screen) were given after the instructions. The subjects were required to speak out aloud the steps that were taken in the evaluation of these queries. The experimenter confirmed correct evaluations and made short comments on errors ('look again', 'are you sure?'). During the experiment no feedback was given about errors.

The stimuli were presented in black on a white 19-inch high-resolution computer screen. The experiment was self-paced; the subject called the next stimulus by clicking on the NEXT-button (see fig. 1). Between the stimuli was a one-second time gap. A warning signal indicated the arrival of next stimulus. Subjects responded by clicking with the mouse on one of the two fields labeled YES and NO. They were allowed to correct their response. All responses were recorded automatically in the computer. In the verbal protocol sessions subjects were asked to think aloud when processing a stimulus. The verbal protocols were recorded on tape.

Results

The latency between presentation of the stimulus and the first response in the speed session was used in further analysis. This response was also analysed for errors.

Latency data

The distribution of the latency data was slightly skewed and showed a long tail with some extreme large values. On the basis of this distribution, latency values greater than 25 s ($n = 27$ out of 3264) were eliminated from the latency dataset. Furthermore, all cases of incorrect response (the errors) were removed. This reduced the dataset from 3264 to 3070 points. The mean latency to process queries was 8.7 with a standard deviation (sd) of 3.5 s.

Analyses of variance were performed on the data for the different logical operators and their combinations. The two subsets of stimuli used in the speed session were grouped together, because no differences were found. Mean latencies for the four query categories, single, double, mixed, mixed( ), crossed with the three levels of NOT are presented in table 2. The difference between the single and double category is a reflection of the extra processing time of a 3-clause versus a 2-clause query. No significant difference was found in latencies to process single and double queries containing AND operator(s) versus those containing OR (means 8.2 s and 8.1 s, respectively). The differences between double and mixed, and double and mixed( ) were not significant. The latencies of queries with parentheses ('mixed( )') showed a

Table 2
Average response latencies (s) for the four query and three NOT categories.

	Query categories				
	Single	Double	Mixed	Mixed( )	Mean (*sd*)
0-NOT	6.2	8.3	8.9	8.3	7.7 (3.2)
1-NOT	7.0	9.4	10.3	8.9	9.1 (3.7)
2-NOT	8.1	9.9	10.7	9.1	9.5 (3.9)
Mean (*sd*)	6.9 (3.0)	9.1 (3.3)	10.1 (3.9)	8.8 (3.5)	8.7 (3.5)

significant improvement compared to those without them ('mixed'), ($F(1, 1443) = 34.8$, $p < 0.001$). An analysis of the position of AND–OR in the queries of the mixed categories indicated no order effect.

The effect of number of NOTs was significant ($F(2, 3064) = 57.6$, $p < 0.001$). A post-hoc test showed that the effect was due to the difference in latencies between 0-NOT and 1-NOT ($F(1, 3064) = 58.8$, $p < 0.001$). No significant higher order interactions were found for the query categories and the NOT's.

Visual scanning of the data showed similar performance of the subjects across different logical combinations, although latencies differed from subject to subject. In particular the university students processed the queries faster than the high school students. For the two groups of students an analysis of variance across the four query categories and NOT categories indicated a significant effect of groups ($F(1, 3062) = 127.6$, $p < 0.001$). No higher order interactions of these factors were found. In a two-way ANOVA a significant effect of blocks and a significant higher order interaction with student groups was found, respectively, $F(7, 3054) = 13.5$, $p < 0.001$; $F(7, 3054) = 3.6$, $p < 0.001$. Latencies dropped over the 8 speed blocks from 10 to 6.6 s for the university students and from 9.9 to 8.9 s for the high school students. The performance of both groups started off the same, but that of the university students improved more than that of the high school students.

Error data

The first main analyses considered the differences between query categories, split out in separate analyses (Pearson chi square) of single vs. double, double vs. mixed, mixed vs. mixed( ). Mean error rates for the four query categories, single, double, mixed, mixed( ), and the levels of the NOT are presented in table 3. The difference between the single and double category is not significant. There is no significant difference in error rates between AND and OR with two- and three-clause queries pooled together. The difference between double and mixed was significant ($\chi^2(1) = 12.2$, $p < 0.001$). The queries with parentheses ('mixed( )') showed significantly fewer errors compared to those without ('mixed'), ($\chi^2(1) = 5.2$, $p < 0.025$). An analysis of the order of AND–OR in the queries of the mixed categories indicated no order effect.

Table 3
Errors (%) for the four query and three NOT categories as percentages of the number of observations (n).

	Query categories				
	Single	Double	Mixed	Mixed( )	Mean (n)
0-NOT	1.6 (384)	2.4 (336)	7.9 (240)	4.2 (192)	3.6 (1152)
1-NOT	7.3 (192)	4.9 (288)	7.3 (288)	3.8 (288)	5.7 (1056)
2-NOT	4.7 (192)	4.9 (288)	8.3 (288)	6.6 (288)	6.2 (1056)
Mean (n)	3.8 (768)	3.9 (912)	7.8 (816)	4.9 (768)	5.1 (3264)

The NOT effect is significant ($\chi^2(1) = 9.2$, $p < 0.005$). A post-hoc test showed that the effect was mainly due to the difference between 0-NOT and 1-NOT ($\chi^2(1) = 5.7$, $p < 0.05$). No significant higher order interactions were found for the query categories and the NOT's.

For the two groups of students a χ^2 across the four query categories and NOT categories indicated no effect of groups. No higher order interactions were found. There was a significant effect for blocks ($\chi^2(7) = 17.7$ $p < 0.025$). The error rate dropped from 4.9% to 1.7% in the last block. However, no significant higher order interaction with student groups was found.

The results of the latency and error data can be summarised as follows (fig. 2):

(i) largest latencies and highest error rates are found in the 'mixed' category; smallest latencies and lowest error rates are found in the 'single' category;
(ii) latencies increase significantly in the 'double' category compared to the 'single', but error rates do not increase;

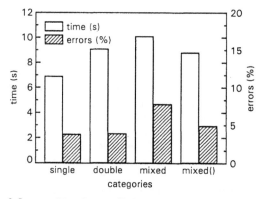

Fig. 2. Latency (s) and error (%) data for the four query categories.

(iii) error rates increase significantly in the 'mixed' category compared to the 'double', but the higher latencies are not significant;

(iv) latencies and error rates decrease significantly in the 'mixed( )' category compared to the 'mixed'.

Verbal protocols

An analysis of the verbal protocol sessions indicated that subjects read the query aloud, processing it from left to right and evaluating clause by clause. Sometimes the whole query was read first; at other times clauses were read and evaluated immediately. The instance information against which the clauses were evaluated was seldom read aloud.

An analysis of the error data from the verbal protocol (VP) blocks (table 4) showed that errors were quite high in the four query categories during the first VP block, especially for the complex queries (AND OR). Average error rate for the first block was 12.3% ($n = 480$), but it reached as high as 22.9% for complex queries with two NOTs. The overall rate dropped to 4.4% in the second verbal protocol block.

Processing errors

To permit a more detailed analysis, the protocols were coded in terms of processing errors and were examined for specific user strategies. Two levels of query processing had been identified in a previous pilot study: the processing of the clauses and the processing of the logical operators using the results of the clauses as input. The following seven categories of processing errors, distributed over the two levels, were observed. They ranged from misreading clauses to incorrectly interpreting the logical structure of the query.

Mistakes concerning the value (truth or falseness) of a clause can be caused by errors in reading the clause, or errors in evaluating the clause (i.e., comparing the clause with the related instance information, sometimes in combination with a NOT). The following four categories contained these cases.

(1) *Reading a clause incorrectly.* Comparators ($=$, $<$, $>$, $< >$) were often misread. For example, subjects read 'smaller than', instead of 'greater than'; or 'is equal to' instead of 'is unequal to'. Also, numerals (e.g. 1963) were misread, especially the numerical values that were used in clauses with 'year' and 'rating'. These misreadings did not always result in an incorrect assessment of the clause. Only those misreadings that did were counted.

(2) *Missing a NOT while reading.* A NOT operator standing before a clause was sometimes missed in the subject's reading of the clause and was therefore not incorporated in further evaluation, resulting in an incorrect assessment of the value of the clause.

(3) *Incorrect evaluation of a clause.* The clause had to be compared to the information given in the LP-instance. Even though a clause was correctly read, subjects could come to an incorrect conclusion about the truth or falseness of the clause after this comparison.

(4) *Incorrect evaluation of a NOT clause.* Wrongly evaluated clauses that were combined with a NOT operator were scored separately in this group.

Table 4
Error data (%) from the verbal protocols of the first and second block for the four query and the three NOT categories as percentages of the number of observations in the categories (n).

Query categories														
	Single			Double			Mixed			Mixed( )			Mean (n)	
Block	1	2	(n)	1	2	(n)	1	2	(n)	1	2	(n)	1	2	(n)
0-NOT	4.2	0.0	(24)	8.3	0.0	(24)	20.8	8.3	(24)	8.3	0.0	(24)	10.4	2.1	(96)
1-NOT	0.0	4.2	(48)	5.6	4.2	(72)	12.5	4.2	(48)	12.5	4.2	(48)	7.4	4.2	(216)
2-NOT	4.2	0.0	(24)	22.9	6.3	(48)	20.8	8.3	(48)	22.9	6.3	(48)	19.6	5.9	(168)
Mean	2.1	2.1	(96)	11.8	4.2	(144)	17.5	6.7	(120)	15.8	4.2	(120)	12.3	4.4	(480)

On the query level, the processing of the logical operators and the combining of the results of the clause assessment are the key activities. The verbal protocols enabled us to keep track of the locus of attention of the subjects while they were processing a query. Errors found at this level are the incorrect omission of a clause and incorrect conclusions about the value of the query.

(5) *Incorrectly omitting a clause.* Sometimes subjects skipped the evaluation of a relevant clause; this was scored as an omission.

(6) *Incorrect evaluation of the query.* In some cases queries whose clauses had been correctly read and evaluated were still processed incorrectly, resulting in an incorrect response. This category contains those errors that could not be assigned to any other.

(7) *Redundant evaluation of a clause.* Most of the time subjects correctly bypassed evaluation of the clauses that were redundant in a query. However, sometimes they did process a redundant clause. Processing of these clauses did not change the logical sense of the query. For example, if the first clause of the query A OR B OR C is found to be TRUE, evaluation of the two other clauses is redundant. This category is not associated with errors in response, and so should be considered separately from the preceding six.

The verbal protocols from the first and last sessions of the experiment were coded using these error categories. For each subject, two analyses were performed: an analysis of the incorrect trials in terms of the processing errors (error categories 1–6); and an analysis of the processing of queries with redundant clauses in the query (error category 7). As a general rule, errors observed in processing were assigned to a category only if the error led to an incorrect evaluation of the query against the instance (except for the redundant clauses, category 7). Table 5 gives the distribution of errors relative to the total number of trials-with-errors for the two verbal protocol blocks. Some incorrect trials contained more than one error count (in the same or different categories). If there was more than one error in the same category in a query evaluation, this was scored as

Table 5
Processing errors in the verbal protocols for the two blocks in percentage of the total number of trials-with-errors (categories 1–6) or of the number of redundant clause queries (category 7).

Error categories	Block 1 ($n = 59$)	Block 2 ($n = 21$)
1. Reading a clause incorrectly	0.0%	14.3%
2. Missing a NOT while reading	49.2%	42.9%
3. Incorrect evaluation of clause	22.0%	4.8%
4. Incorrect evaluation of NOT clause	39.0%	33.3%
5. Incorrectly omitting a clause	32.2%	42.9%
6. Incorrect evaluation of the query	6.8%	4.8%
	($n = 192$)	($n = 192$)
7. Redundant evaluation of a clause	27.6%	7.8%

a single error. Thus, the percentages do not sum to a 100. In block one there were 59 (12.3%) incorrectly assessed queries, and in block two there were 21 (4.4%). There were 16 queries containing redundant clauses.

The data show that the number of errors related to a NOT is relatively high and did not decrease in the second block. The errors related to the redundant evaluation of clauses decreased in the second block. On the other hand, errors associated with incorrectly omitting a clause did not decrease.

Two further details of the processing errors are worth noting. The first concerns the effect of parentheses on the processing of clauses. The presence of parentheses in a query implies that the clauses within the parentheses are to be grouped. Two processing errors are related to this at the query level: the omitting of a clause and the redundant evaluation of a clause (categories 5 and 7). The data show that only the omitting error is affected by the use of parentheses; the error rate was 18.3% and 4.2% for mixed and mixed( ) query category respectively ($n = 120$). This effect was not found for the redundant processing, the means being 23.3% and 25.8% for mixed and mixed( ) respectively ($n = 120$).

The second point concerns the interaction of NOT with the clause comparator. In the processing of a NOT clause either the comparator ($=$, $<$ $>$, $<$, $>$) has to be converted to its opposite sense before clause evaluation; or else the value (true or false) of the clause is converted after clause evaluation. The protocols indicate that, in general the subjects convert the comparator. Conversion of '$=$' is simpler than conversion of '$<$' or '$>$'. A common error was the conversion of 'smaller than' to 'greater than' instead of to 'greater than or equal'. The number of errors was 11.7% ($n = 240$) and 4.5% ($n = 576$) for, respectively, 'NOT $<$' or 'NOT $>$' and 'NOT $=$'. Logically equivalent notations 'NOT $=$' and '$<$ $>$' differed in number of errors, respectively, 4.5% ($n = 576$) versus 1.3% ($n = 312$).

Table 6
A sample stimulus with a typical protocol (A) and a protocol that reveals the mechanisms of the logical operators (B).

---

not year $=1979$ and (genre $=$ jazz or artist $<$ $>$ Jody Sylvian)
Will Bazar   classical   1965   1

(A) 'not year is 1979...correct...and the genre is jazz...not correct artist is not JS, that one is correct...'
    (subject mq, 2nd block, answer is correct)

(B) 'in this case we begin with an "and" and then an OR between parentheses...the year must not be 1979...that is satisfied because it is 1965...look to the other side of the 'and'...we can choose...either the genre has to be jazz...it's not...or the artist has to be unequal to JS but that is true, thus both sides of the "and" are satisfied by the conditions and the answer should be yes'.
    (subject ea, 2nd block, answer is correct)

User strategies

In general subjects processed the queries from left to right even when parentheses were present in the query. Only 1.7% ($n = 720$) of the queries were approached in a different order. Typically, protocols revealed the sequence of clause evaluation, but did not reveal directly the understanding of the logic in the query. They had the form 'evaluate A, evaluate B, (evaluate C), state conclusion'. There are some protocols, however, that show in detail the mechanisms of the logical operators. Table 6 shows samples of two protocols.

The verbal protocol A shows a fairly short protocol in which almost no references are made to logical operators or the organisation of the processing. In verbal protocol B explicit reference was made to the mechanisms of the operators and the organisation of the query. The protocol type A was the typical protocol.

In summary, the verbal protocols showed that, in general, queries were processed from left to right. Parentheses did not change the order of processing. A large proportion of the errors were due to reading errors, most notably in relation to NOT clauses. Also the flipping of the greater or smaller than sign was a major source of errors. The unequal comparator ('$<$ $>$') caused less problems than the logical equivalent 'NOT $=$'. The total number of errors in the second verbal protocol block was less than half that in the first block.

Discussion

The latency and error data converge to show that subjects have increasing difficulty with longer, more complex queries: queries involving mixed (AND OR) and NOT operators were the most difficult to interpret. The inclusion of parentheses in the query had a strong effect on latencies and errors. An increase in the number of logical operators from one to two had no effect on the number of errors. No differences between simple processing of AND and OR were found, in contrast with the experimental findings of Greene et al. (1990).

To interpret these findings we consider first the verbal protocol data and then develop a model aimed at identifying latency and error components in the query processing.

The verbal protocols gave insight into certain aspects of the query processing. Most importantly, they revealed how the subjects organised their analysis of the query. The verbal protocols showed that queries were always processed from left to right, including those cases in which it would have been more efficient to start at the end of the query. For instance, the query

A AND B OR C

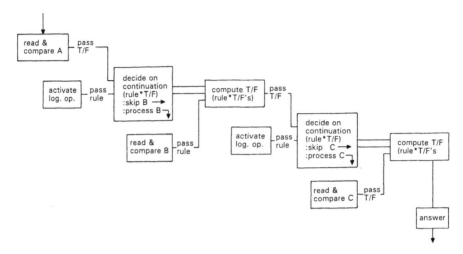

Fig. 3. Model of query processing of a three clause query. (A, B, C: the clauses; log.op.: logical operator; T/F: true/false).

can be analysed by identifying the part of the query that has the greatest impact on the truthness or falseness of the whole query. Here, an efficient strategy would be to evaluate C first, because if C is true then the whole query is true and the answer for the query can immediately be given. This hierarchical approach was, however, not used by the subjects.

A model of query processing can be formulated in which the basic processes are *reading* the elements of the query (the clauses and the logical operators); *comparing* clauses against the instance; *storing* the resulting values of the clauses; and *activating* the rule of the logical operator by retrieving it from memory and storing for immediate use. In addition to making provision for application of the logical operator to the truthness or falseness of the combined clauses, the model must also dictate the flow of processing. For instance, in A AND B OR C, clause B can be skipped if A is false. Thus, two other processes are: *deciding* how to go on and *computing* the value of the logical combination of the clause values.

In fig. 3 this model of query processing is presented for a three clause query. The flow is from left to right. The first step is the processing of clause A, comprised of the basic processes of *read and compare* against the instance. The resulting TRUE/FALSE value of clause A is passed on to the *decide* block, which also requires the rule for the logical operator following the clause. The rule is obtained from the block below, *activate* log.op. The decision on how to continue depends on the combination of logical rule and value of clause A. For instance, if the clause is true and the logical operator is AND, then clause B must be processed; if the clause is false, B can be skipped.

In either case, *compute* produces a temporary value based on the combination of logical rule and values passed from processing of clause A and B (if done). If B is skipped, then the value of clause A alone is passed as the value of the query so far. At this stage, the next logical operator must be *activated* to *decide* how to continue. Here the same thing happens as at the former decision point. If clause C can be skipped, *compute* returns the value of the query based on clauses A and B and the answer can be given.

Given that the processing is from left to right, it is possible to specify which elements in the query must be successively processed to arrive at a conclusion on the query. In table 7, these clauses and operators are specified from left to right for all the possible TRUE/FALSE combinations. Note that no precedence rules were used in this experiment. We assumed that subjects with no programming knowledge or specific training on logic would also lack knowledge of specific rules. The precedence rules would add an additional degree of complexity to the query processing. In fact, only the query A OR B AND C with clause A being TRUE might have been problematic, because, using the precedence rule, it has to be processed differently.

From table 7 it can be seen that with some combinations of logical operators and TRUE/FALSE values in the mixed category, more clauses need to be processed than in the other categories. In most cases, both logical operators have to be processed to arrive at a conclusion about the value of the query. In some cases in the mixed( ) condition, however, the value of the query can be computed after processing only the first operator.

From the processing specifications in table 7 it can be concluded that the average amount of processing will be different for the different query categories. Now, if we assume that the two main processing components are the processing of clauses and the logical analysis (activate, decide, compute) and that these two components cost the same amount of time, we can use the number of processing components as a rough estimate of the latencies. Totalling the processes, the

Table 7

The clauses (A, B, C) that have to be processed to determine the value of the query for the different values of the clauses (T = true, F = false) given a left to right processing of the query. The logical operator to be processed is indicated by $\circ$. (NB: no precedence rules were used in the experiment.)

Query categories

Double						Mixed						Mixed( )					
AND AND			OR OR			AND OR			OR AND			AND (OR)			OR (AND)		
Clause																	
A	B	C	A	B	C	A	B	C	A	B	C	A	B	C	A	B	C
T $\circ$	T $\circ$	T/F	F $\circ$	F $\circ$	T/F	T $\circ$	F $\circ$	T/F	F $\circ$	T $\circ$	T/F	T $\circ$	F $\circ$	T/F	F $\circ$	T $\circ$	T/F
T $\circ$	F $\circ$		F $\circ$	T $\circ$		T $\circ$	T $\circ$		F $\circ$	F $\circ$		T $\circ$	T $\circ$		F $\circ$	F $\circ$	
F $\circ$	F $\circ$		T $\circ$		$\circ$	F $\circ$		T/F	T $\circ$		T/F	F $\circ$			T $\circ$		

Table 8
Predicted and (between parentheses) observed average response latencies (s) with the Double data as basis.

	Query categories		
	Double	Mixed	Mixed( )
0-NOT	(8.3)	8.9 (8.9)	7.6 (8.3)
1-NOT	(9.4)	10.2 (10.3)	8.6 (8.9)
2-NOT	(9.9)	10.7 (10.7)	9.1 (9.1)

ratio of amount of processing in the three conditions is, on average, $1:1.08:0.92$ for double:mixed:mixed( ). Other time aspects, like e.g. start and answer, are considered to be negligibly small compared to the two major time components. Thus, on the basis of the model and processing characteristics, the latencies in the mixed and mixed( ) categories can be predicted using the double category as norm. The predicted and the observed latencies are presented in table 8. Each level of NOT was considered as a separate case, because we do not see yet how that factor can be incorporated in the model. The predictions were confined to queries with equal number of operators (three). Extrapolation to the single operator queries showed underestimation of processing time.

This analysis shows that the effect of parentheses is to reduce the amount of processing needed in the mixed( ) category, since processing of the second operator can sometimes be omitted. It is not clear precisely what further effect the parentheses might have on processing; further experiments investigating this aspect should control for the number of clauses processed in the parenthesis and non-parenthesis cases.

These comparisons show that latencies can be predicted with some accuracy. The intention in using this model is to show that one way to explain the latency results is on the basis of the number of processing components in the query. Further research is needed to investigate the assumption that the time for the different processes are equal.

How can the model help to interpret the error data? From the protocol data we saw that the number of reading errors was relatively high in proportion to other errors. Reading errors occur during the processing of the clauses, in particular in combination with the NOT operator. Following the same line of reasoning as with the latency data, the number of clauses to be processed (from table 7) could be used to predict the error data in the speed trials. The ratio of the number of clauses processed in the three conditions is thus computed as $1:1.33:1$. However, the error rate predicted using these assumptions differs from the observed data. The fact that the double and single conditions did not differ in number of errors (despite a two-fold difference in number of clauses processed) also argues against using amount of processing as a basis for error prediction (see fig. 2). A significant difference between double and mixed queries lies in the fact that the same logical rule is invoked in the double case whereas two different rules are invoked in the mixed case. Interference between the two different rules might explain the rise of errors in the mixed condition. The parentheses in mixed( ) might have helped the subject to mentally segregate the processing of the two operators, resulting in a large reduction of errors. It is unclear which factor plays a role in the interpretation of queries containing parentheses.

A somewhat surprising result concerned differences between university and high school students. Subjects were selected as having no specific logical training or programming experience. No a priori differences were expected between the university and high school students. Indeed, both groups started off at the same processing speed and did not differ in the error data. Both groups improved performance in term of errors. However, the university students became relatively faster in the successive blocks. In terms of speed–accuracy tradeoff their performance became more efficient than that of the high school students. Both groups learned during the experiment although there was no feedback given during the experiment.

A remark is necessary concerning the methodology. The combination of latency and error measurement and verbal protocols was a good methodology for addressing the problem of the interpretation logical operators. The protocols gave insight into details of the query processing. However, the application of the logical rules was not explicitly referred to in the protocols. One plausible explanation for this is that the problem the subjects had to solve did not evoke deliberate problem solving, because the rules were too simple and were therefore processed quickly. When processes are fast, fewer verbal references are made to those processes (Ericsson and Simon 1984).

Most studies have addressed query production, i.e. the creation of a query. In our study we studied a process that is assumed to be part of query production, namely, query interpretation. The reason for selection of this strategy was that the understanding of logical operators should be separated from other factors in querying production, like knowledge of the dataset or understanding of the data model. We regard query interpretation as a prerequisite for query production. Understanding logical operators is necessary for the active use of them. The difference between these two task situations is indicative of the extra processes that apply in query formulation investigated in other studies. Our results indicate that people can successfully interpret logical operators. However, the subjects did not make use of efficient strategies to solve the problems. The number of errors was low in this task situation compared to the results from studies in query generation. (Greene et al. (1990) reported an error rate of 47.5% in the generation of SQL-type of queries with AND OR operators.) In this study we have used relative simple problems. More complex problems could force people to explicitly organise the processing of the query. Further studies on logical operators should address that organisation process and the question of how to support it.

Appendix

Table A.1
Queries in the verbal protocol blocks (the first block equals the second block).

	Query categories	
	Single	Double
0-NOT	= and > < > or < >	= and > and = = or = or <
1-NOT	not = and > = and not = not = or = < or not =	not < and = and = < > and not < and = = and < > and not = not < > or = or = = or not < or < > < or = or not >
2-NOT	not = and not = not = or not =	not = and not = and = not = and < > and not > not = or not > or = = or not > or not <

	Query categories	
	Mixed	Mixed( )
0-NOT	= and < > or > > or = and >	= and (= or =) < > or (= and < >)
1-NOT	not = and = or < > < > and not = or > < > or not = and = = or < > and not >	not = and (= or < >) < and (= or not =) not > or (= and =) = or (= and not =)
2-NOT	not = and not = or = = and not = or not = not > or not = and < > not < or = and not >	not = and (not = or >) > and (not = or not =) not = or (not = and =) = or (not = and not =)

References

Borgman, C.L., 1986. The user's mental model of an information retrieval system: An experiment on a prototype online catalog. International Journal of Man–Machine Studies 24, 47–64.

Broadbent, D.E. and M.H.P. Broadbent, 1978. The allocation of descriptor terms by individuals in a simulated retrieval system. Ergonomics 21, 343–354.

Brosey, M. and B. Shneiderman, 1978. Two experimental comparisons of relational and hierarchical database models. International Journal of Man–Machine Studies 10, 625–637.

Chamberlin, D.D., M.M. Astrahan, K.P. Eswaran, P.P. Griffiths, R.A. Lorie, J.W. Mehl, P. Reisner and B.W. Wade, 1976. SEQUEL 2: A unified approach to data definition, manipulation and control. IBM Journal of Research and Development 20, 560–575.

Cooper, W., 1988. Getting beyond Boole. Information Processing & Management 24, 243–248.

Ericsson, K.A. and H.A. Simon, 1984. Protocol analysis: Verbal reports as data. Cambridge, MA: MIT Press.

Greene, S.L., S.J. Devlin, P.E. Cannata and L.M. Gomez, 1990. No IFs, ANDs, or ORs: A study of database querying. International Journal of Man–Machine Studies 32, 303–326.

Jarke, M. and Y. Vassiliou, 1985. A framework for choosing a database query language. Computing Surveys 17, 313–340.

Katzeff, C., 1986. Dealing with a database query language in a new situation. International Journal of Man–Machine Studies 25, 1–17.

Lochovsky, F.H. and D.C. Tsichritzis, 1977. User performance considerations in DBMS selection. Proceedings of ACM SIGMOD. New York: Association for Computing Machinery. pp. 124–134.

Marchionini, G., 1989. Making the transition from print to electronic encyclopaedias: adaptation of mental models. International Journal of Man–Machine Studies 30, 591–618.

Michard, A., 1982. Graphical presentation of boolean expressions in a database query language: design notes and an ergonomic evaluation. Behaviour and Information Technology 1, 279–288.

Ogden, W. and C. Kaplan, 1986. The use of 'and' and 'or' in a natural language computer interface. Proceedings of the Human Factors Society 30th Annual Meeting. Santa Monica, CA: Human Factors Society. pp. 829–833.

Reisner, P., 1977. Use of psychological experimentation as an aid to development of a query language. IEEE Transactions on Software Engineering SE-3, 218–229.

Reisner, P., R.F. Boyce and D.D. Chamberlin, 1975. Human factors evaluation of two data base query languages – Square and sequel. Proceedings of the National Computer Conference. Arlington: AFIPS Press. pp. 447–452.

Thomas, J.C. and J.D. Gould, 1975. A psychological study of query by example. Proceedings of the National Computer Conference. Arlington: AFIPS Press. pp. 449–445.

Thompson, R.H. and W.B. Croft, 1989. Support for browsing in an intelligent text retrieval system. International Journal of Man–Machine Studies 30, 639–668.

Vassiliou, Y. and M. Jarke, 1984. 'Query languages – A taxonomy'. In: Y. Vassiliou (ed.), Human factors and interactive computer systems. Norwood, NJ: Ablex. pp. 47–82.

Wason, P.C. and P.N. Johnson-Laird, 1972. Psychology of reasoning: Structure and content. Cambridge, MA: Harvard University Press.

Welty, C., 1985. Correcting user errors in SQL. International Journal of Man–Machine Studies 22, 463–477.

Acta Psychologica 78 (1991) 227–241
North-Holland

Analysing the deep structure of queries: Transfer effect on learning a query language *

Lena Linde and Monica Bergström

University of Stockholm, Stockholm, Sweden

This experiment was designed to investigate the impact on performance, while learning a query language, of specific instructions pertaining to the 'deep-structure' of the queries. The instructions related to deep-structure comprised training in analysing queries in natural language in terms of logical constituents. The effect of such instructions was investigated for two different query languages. One was a textual and one was a 'graphic' query language. The results suggested that the usefulness of explicit training in analysing queries in terms of logical deep-structure is dependent on the semantic constraints in the query language. The textual query language, which had keywords and a syntactical template such as GET_WHERE_ EQ_ assumedly induced analysis of the query regardless of pretraining, while the graphic used in this study tended to encourage a trial-and-error approach, especially when no prior training in analysing queries was given.

Introduction

The motive for the present study was the idea that understanding and mastering of complex tasks, such as formal query writing in database search, is facilitated by explicit instructions pertaining to the so-called deep structure of the task. According to Craik and Lockhart (1972) the processing of verbal information may vary in depth. One end of the depth dimension is represented by the perceptual properties of the verbal material (the surface-structure) and the other end by the semantic properties (the deep-structure). Craik and Lockhart also claimed that depth of processing is related to familiarity with the linguistic material (the more familiar the material, the deeper the level

---

* Requests for reprints should be sent to L. Linde, National Defense Research Establishment, Dept. 5, S-17290 Sundbyberg, Sweden

of processing). Linde and Bergström (1988, 1990) have demonstrated a relationship between ease of learning basic search principles in a relational database and prior knowledge about its informational content and organization. One way to interpret those results is that prior knowledge of the informational content and organization affected the depth at which the subjects processed the written queries. For example, knowledge of informational content and organization may have facilitated the creation of mental models pertaining to the logical aspects of the queries. (To reformulate queries in set-theoretical terms with Boolean operators may be an essential part of information-seeking in a relational database.) Rist (1989) has argued for a distinction between surface- and deep-structure of a program. According to him the surface-structure comprises of a template of the program code (e.g. input and output functions), whereas the deep-structure consists of goals, subgoals and means of the task (what must be accomplished?). Rist showed that novice programmers' creation of plans showed a top-down development (from goal via subgoals to program code). However, once the programmers had learned complete plans, their program creation showed bottom-up development, that is the program code was constructed in the order, in which it occured in the final programs. The study of Rist indicates that comprehension of a programming task at a basic level in terms of goals and subgoals is an essential part of the development of expertise in the programming task.

The study of Rist addresses the question of how level of expertise affects the order, in which relevant knowledge is accessed in a programming task. Adelson (1981), on the other hand, has addressed the question of how level of programming expertise affects the structure of the mental representation of knowledge. She used a recall of randomly presented program lines procedure to infer that novices tended to use a syntax-based organization, that is the surface-structure of a program, whereas experts used a hierarchical organization based on the function of the program, i.e. its deep-structure.

The purpose of the present study is to investigate the effect of giving explicit training in observing the 'deep-structure' of queries to relational databases. It is hypothezised that such training facilitates subsequent learning of the formal rules of a query language. The deep-structure of a query is defined as the result of analysing a query written in normal Swedish in terms of 'what is asked for' and 'what is known'. These parts constitute information to be retrieved and condition(s). A condition consists of property (a column label, e.g. name), specific value (e.g. Bob) and a relational operator. The surface-structure of a query, on the other hand, was defined as the end-product of translating a query into a formal query language. Accordingly, the surface-structure of a query varies over different formal query languages, whereas the deep-structure is the same. It was assumed that the acquisition of the formal syntax of a query language, which from the user's point of view may appear arbitrary, would be facilitated by prior knowledge of the underlying structure of the task. Two different query languages were used to investigate this hypothesis. One of the query languages was textual and required that commands to the system were typed on an alpha-numeric keyboard. The other was graphic. Commands to the system in this query language were given by pointing at particular places in a graph with a lightpen. The aim of using two quite different query languages was to check the generalizability of the hypothezised effect of deep-structure knowledge. In addition to query formulation tasks, subjects were given some simple programming tasks. These tasks comprised writing 'procedures', which could be used to retrieve information from the database without having to formulate queries in any of the formal query languages. The purpose of giving the programming task was to investigate possible transfer effects of knowledge related to deep-structure of queries.

Method

Subjects

Twenty subjects volunteered to take part in the experiment. They were all recruited from a three-year, undergraduate education in behavioral sciences. The subjects were paid 250 SEK for their participation. Twelve subjects were female and eight were male (two male and three female under each experimental condition). None of the subjects had any basic training in computer science or in computer languages such as PASCAL or BASIC.

Design

The subjects were randomly assigned to one of four experimental conditions according to a two-factorial design (see table 1). One factor was pretraining in analysing queries in written Swedish in terms of information to be retrieved and logical conditions vs. no such training. The other factor comprised type of query language,

Table 1
Experimental design.

	Query language	
	Textual	Graphic
Deep-structure training		
Yes	5 subjects	5 subjects
No	5 subjects	5 subjects

which should be learnt. One was the textual query language (see Appendix). The other was the graphic query language. Fig. 1 shows the layout of the graph, on which the subjects should point with a lightpen. [1] The graph was horizontally placed on a table to

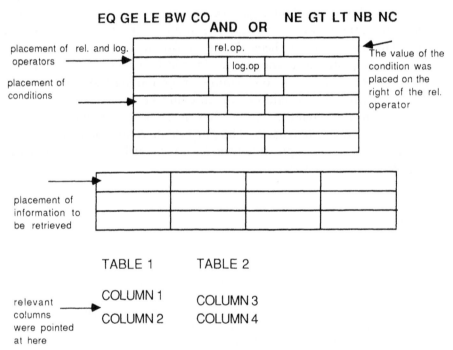

Fig. 1. Layout of the table used for the graphic query language. Text in capital letters is part of the ordinary layout, that was presented to the subjects. Text in small letters is given here for explanatory purposes and was not presented to the subjects.

[1] The graphic query language is an experimental query language constructed by Mats Lind, Bengt Göransson and Patrik Schwalbe at the Computer Center (UDAC) in Uppsala, Sweden. The hard- and software environment in which it was run in the present study was not ideal with respect to system response time. The program was not used in the present study for the purpose of evaluating its user-friendliness.

the right of a VDU. All text in capital letters in fig. 1 is part of the ordinary layout of the graphic query language. Text in small letters in fig. 1 is not part of the ordinary layout. It is only given here to convey information to the reader.

Manuals and oral instructions

Pretraining in analysing textually represented search tasks in terms of basic query components was given in the form of (a) a table containing three columns and six rows, and (b) four queries pertaining to information in the table. The subjects were asked to analyse each query in terms of 'what is asked for' (i.e. information to be retrieved and 'what is known' (i.e. information constituting condition). Furthermore, they were instructed to analyse the condition part in terms of property and value of the condition.

Subjects presented with the textual query language were given a manual containing information about basic concepts (definitions), such as keywords, relational and logical operators, syntactical rules and query templates. Subjects presented with the graphic query language were given a manual containing information about how and where to point at the graph presented in fig. 1, about the meaning of conditions, relational and logical operators and about how and where to use the keyboard for typing the values of conditions. Instructions pertaining to conditions and operators were the same for both query languages. Both manuals were supposed to contain necessary information, but no more, for accomplishing the tasks.

All subjects were also given an overview of the data-organization, as shown in fig. 2. A 1-page description of a scenario, pertaining to the contents of the tables was also given to all subjects.

The manual for the second part of the experiment comprised a description of different modes (edit mode etc.) in the MIMER system (system used in the present study), of some relevant edit commands, instructions about three types of statements in the special-purpose programming language and some general information about what purposes procedures might be serving. The three types of statements were (a) defining and assigning values to variables, (b) execution of queries containing variables, and (c) constructing a menu option, by which a user could gain access to a certain procedure.

The subjects' were also given oral instructions by the experimenter. These instructions included a summary of the contents of the manual and a demonstration of its organization (given to all subjects). The experimenter also gave assistance when a subject was found to be stuck during the performance of a task either by referring to a specific section of the manual or by informing the subject about the nature of the error.

ACTIVITIES				PERSONS		GROUPS		
TIME	NAME	PLACE	ACTIVITY	NAME	PROFESSION	PROFESSION	PLACE	TIME
745	Jim	x-town	road-blocking	Jim	policeman	policemen	x-town	745
(92 rows)				(25 rows)		(4 rows)		

Fig. 2. Overview of the data organization, as presented to the subjects.

The subjects' tasks

The experiment had two parts. One comprised the learning of one of the two query languages mentioned above. The second part comprised the learning of a simple, special-purpose programming language. The second part was equal for all subjects.

The first part contained 10 search tasks (queries formulated in Swedish) including one or two conditions (combined by AND, OR, NOT respectively), the relational operators EQUAL or LESS THAN and queries, in which data from one or two columns had to be retrieved. The subjects performed the search tasks interactively at a VDU.

The second part comprised four procedure construction tasks (see Appendix): (a) Writing a procedure, by which a user could execute a specific, formal query repeatedly by one EXECution command (one-line program code). (b) Writing a procedure, by which a user could execute a query having a constant 'information to be retrieved' constituent and a variable condition (two-line program code). (c) Writing a procedure, by which a user could execute a query having both a variable 'information to be retrieved' constituent and a variable condition (three lines program code). (d) Constructing a procedure, which allowed a user to choose between one of the three procedures mentioned above from a menu (three-line program code). The subjects were instructed to first write the program codes on paper then enter the codes to the system, test them and correct them (if necessary). A line-oriented editor had to be used to enter and correct program code.

Experimental procedure

A relational database system (MIMER) was used throughout the experiment.

Each subject took part in the experiment individually on two occasions (separated by at least one day and at most one week). Every subject worked 4 hrs on each occasion (see fig. 3). The first session comprised training in formulating specific queries in one of the two types of query languages. Subjects under the pretraining on deep-structure condition began this session with the pretraining task. The other subjects began the session with instructions about the query language. The second session comprised training in writing 'procedures' using the special programming language.

Each 4-hour session ran as follows: First, the experimenter gave an oral presentation of the manual content and its organization. Then the subject could read the manual silently as long as (s)he wanted. After that the experimenter posed questions about essentials of the manual (such as definitions of the technical concepts). The presentation and reading of the manual, and questioning took between one and two hours. Then the subjects performed as many of the training tasks (10 query construction tasks on the first session and 4 procedure construction tasks on the second session) as they could manage interactively at a VDV using the manual. The subjects were allowed to ask the experimenter for help when necessary. The experimenter could also provide verbal assistance, when the subject appeared to be 'stuck'. The query construction tasks were presented in the same order to all subjects and each task had to be successfully completed before proceeding to the next. As for the procedure construc-

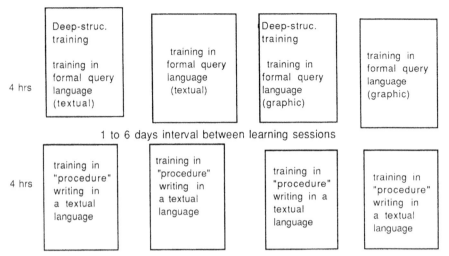

4 hrs

| Deep-struc.
training

training in
formal query
language
(textual) | training in
formal query
language
(textual) | Deep-struc.
training

training in
formal query
language
(graphic) | training in
formal query
language
(graphic) |

1 to 6 days interval between learning sessions

4 hrs

| training in
"procedure"
writing in
a textual
language | training in
"procedure"
writing in
a textual
language | training in
"procedure"
writing in a
textual
language | training in
"procedure"
writing in
a textual
language |

Fig. 3. Experimental procedure under the four conditions.

tion task the criterion of a successful solution was that the procedure should be in working order. All procedure codes had to be written on paper before any of them were entered to the system but the subjects were free to enter the program code of a new procedure before testing a previous one.

Collection and analyses of data

Log-files of transactions (attempts to construct queries, entering or editing program statements in the 'procedures') were collected. Parts of the verbal communication between the experimenter and subject were tape-recorded and typed, all of which were task-relevant dialogs between experimenter and subject.

All errors (with the exclusion of typing and spelling errors) were noted by one of the authors using 14 categories for the query construction task and 25 categories for the procedure construction task on the basis of log-files and verbal protocols (The categories are shown in the Appendix). 'Assistance' given by the experimenter and so-called self-corrections of errors (errors spontaneously corrected by the subject without assistance from the experimenter) were identified on the basis of the verbal protocols by the same author.

The cause of every error during query construction was scored by two judges (the authors) independently using six categories: (1) Misunderstanding of oral and written instructions. (The judges used the criterion that the error should pertain to information given in immediate connection with its appearance.) (2) Lack of attention. (The judges used the criterion that the error should be corrected by the subject spontaneously, without interference from the experimenter, or that an error should appear in a previously correct part of a query.) (3) Disregard of syntactic constraints. (The judges

Table 2

Day 1 (query construction): Means (and standard deviations) of no. of errors (not self-corrected) and frequency of assistance. All figures are corrected for number of tasks solved (see text). PADS = Pretraining in Analysing Deep-Structure of queries; TQL = Textual Query Language; GQL = Graphic Query Language. (Efficiency = average no. of completed tasks.)

	Efficiency	Errors		'Assistance'	
		M	(SD)	M	(SD)
TQL					
PADS	7.4	10.6	(10.2)	11.7	(12.5)
no-PADS	8.0	6.7	(5.2)	6.1	(6.3)
GQL					
PADS	6.8	4.6	(2.3)	6.3	(4.6)
no-PADS	6.8	19.1	(10.4)	23.5	(13.6)

used the criterion that the subject should seem unaware of the necessity to strictly adhere to formal rules of the query language.) (4) Inability to analyse query properly. (The judges used the criterion that the subject should appear to be unable to understand the deep-structure of the query.) (5) Disregard of data organization. (The judges used the criterion that the subject should use an irrelevant table, or should associate a table label with a column label, which did not exist in the table in question.) (6) Others, (Retrieval of redundant information or an inefficient but correct solution of task).

Results

Performance on the query construction task (day 1)

Table 2 shows the average number of errors (with the exclusion of errors that were spontaneously corrected by the subject without assistance from the experimenter, that is 'self-corrections') and assistances under the four experimental conditions. Each number is corrected for the number of completed tasks according to the formula;

$$N_{total}/N_{completed} \times X_i,$$

where N_{total} is 10, $N_{completed}$ the number of completed tasks of each subject and X_i number of errors or frequency of assistance of each subject. The left column shows the average 'efficiency', i.e. number of completed search tasks, under different experimental conditions. The efficiency differences between experimental conditions are small. As to both number of errors and assistances given, the condition *graphic query language / no training on analysing deep-structure* exhibits the lowest performance and the condition *graphic query language / training in analysing deep-structure* the highest performance. Two-way univariate analyses of variance with type of query language on day 1 as one

Table 3

The proportions of errors of different types (see text) under each experimental condition; 1 = misunderstanding of oral and written instructions; 2 = lack of attention; 3 = disregard of syntactic constraints; 4 = inability to understand query properly; 5 = disregard of data organization; 6 = others. Initial = classifications, about which the judges independently of each other had reached a common decision. Final = classifications agreed upon after a discussion and unanimous decision.

Error cat.	TQL				GQL			
	PADS		no-PADS		PADS		no-PADS	
	Initial	Final	Initial	Final	Initial	Final	Initial	Final
1	0.06	0.04	0.10	0.05	0	0	0.05	0.05
2	0.41	0.44	0.15	0.33	0.40	0.50	0.23	0.13
3	0.21	0.19	0.25	0.23	0.25	0.21	0.28	0.34
4	0.24	0.21	0.30	0.23	0.20	0.17	0.31	0.35
5	0.06	0.06	0.15	0.10	0.15	0.13	0.05	0.05
6	0.03	0.06	0.05	0.05	0	0	0.08	0.07
Total no. categor. errors	34	52	20	39	20	24	39	77

factor and training on deep-structure vs. no training as the other factor gave a significant Interaction between the two factors (no. of errors: $F(1, 16) = 6.91$, $p < 0.05$; frequency of assistance: $F(1, 16) = 6.52$, $p < 0.05$; no. of errors and frequency of assistance: $F(1, 16) = 7.38$, $p < 0.05$). The overall effects of type of query language or training on deep structure vs. no such training were not significant. One-way analyses of variance were also made separately for each query language. Training in deep-structure had a non-significant effect when the query language was textual, but yielded a significant effect on no. of errors + frequency of assistance, $F(1, 8) = 9.58$, $p < 0.05$, when the graphic query language was taught.

Table 3 shows the proportions of errors, scored in terms of six different 'causes' (i.e. all solution attempts noted as an error by one of the authors in terms of any of the 14 categories shown in the Appendix). Initial categorizations are those causes which both judges agreed upon from independent scorings, and final are all causes for which the judges reached an agreement after a discussion and mutual decision. (The judges made a mutual decision for all errors that they did not agree upon independently of each other.) Interrater agreement coefficients were computed as;

$$E_x/E_t,$$

where E_x is no. of error causes agreed upon initially and E_t the total no. of solution attempts initially scored as an error by one of the authors. The interrater-agreement coefficients computed in this way for each experimental condition were 0.65 (TQL/PADS), 0.51 (TQL/no-PADS), 0.83 (GQL/PADS) and 0.51 (GQL/no-PADS), respectively. The two *no training on analysing deep-structure* conditions give somewhat

Table 4
Day 2 (procedure construction): Means (and standard deviations) of no. of errors and frequency of 'assistance' for each experimental condition. All figures are corrected for no. of solved tasks (see text).

| | Writing program code on paper | | | | Entering program code on terminal | | | |
| | Errors | | Assistance | | Errors | | Assistance | |
	M	(SD)	M	(SD)	M	(SD)	M	(SD)
TQL								
PADS	6.4	(3.6)	7.4	(8.0)	12.0	(5.7)	12.0	(7.0)
no-PADS	3.6	(2.3)	6.0	(3.7)	12.6	(5.5)	10.3	(6.6)
GQL								
PADS	6.1	(3.3)	7.1	(5.7)	15.9	(9.6)	7.7	(6.2)
no-PADS	8.6	(1.6)	8.6	(3.5)	12.8	(5.9)	9.6	(4.1)

lower interrater-aggreement coefficients than the other two experimental conditions. This fact suggests that it might have been more difficult to classify the causes of errors committed by subjects not given training on analysing deep-structure.

Performance on the procedure construction task (day 2)

Table 4 gives the number of errors committed and the frequency of assistance given during the procedure construction task. All subjects had been instructed first to write the program code for each procedure on paper and then enter them to the system. The table gives errors committed and frequency of assistance received during these two phases separately. All numbers are corrected for number of completed tasks according to the same formula as for the performance on day 1. The largest number of errors during the 'writing on paper' phase is committed by subjects under the condition *graphic query language / no training on analysing deep-structure* and the smallest number by subjects under the condition *textual query language / no training on analysing deep-structure* condition. The type of query language or training on deep-structure vs. no such training on day 1 had no significant effects on learning the procedure construction task on day 2. The interactions between the two factors were also non-significant. However, the interaction between type of query language and training on deep-structure vs. no such training on day 1 approached a significant level with respect to no. of errors committed while writing program code on paper: $F(1, 16) = 4.36$, $p = 0.053$.

Table 5 shows proportions of errors initially scored by one of the authors as being of different types. (The scoring was based on log-files and verbal protocols.) The four categories are based on mergings of 14 of the 25 categories shown in the Appendix.

Correlations between dependent variables

Table 6 shows partial correlations between different dependent variables (computed after extraction of variance attributable to query language effect, training effect or

Table 5
Proportions of errors (scored by one of the authors) of different types during day 2 (procedure construction).

Type of error	TQL		GQL	
	PADS	no-PADS	PADS	no-PADS
1. Variable definition	0.02	0.10	0	0.08
2. Formulation of prompts and other guiding text for the user	0.49	0.32	0.40	0.31
3. Mismatch between transaction and system mode	0.28	0.41	0.47	0.43
4. Improper use of editor	0.21	0.17	0.13	0.18
Total no. of the above types of errors	43	47	45	52

interaction). The partial correlation between no. of errors and assistance frequency of on day 1 is $+0.89$, whereas the partial correlations between no. of error and frequency of assistance on day 2 are $+0.40$. The procedure construction task on day 2 consisted of two subtasks: writing program code on paper and entering the code to the system. The minimum standard that should be met by the program code on paper had to be set rather low. Some subjects' solution attempts on paper were far from perfect, but they would nevertheless not have benefitted from more assistance at that stage. Consequently they needed more help while at the terminal, which shows in the $+0.60$ partial correlation between errors on paper and frequency of assistance at terminal. One may

Table 6
Partial correlations between different performance variables (after extraction of variance attributable to type of query language, pretraining on deep-structure vs. no such training and interaction). (E = errors, A = assistance.)

	1	2	3	4	5	6
1. Errors, day 1		0.89	0.31	0.02	0.45	0.29
2. Assistances, day 1			0.31	0.01	0.42	0.34
3. E, paper, day 2				0.23	0.40	0.60
4. E, terminal, day 2					0.18	0.40
5. A, paper, day 2						0.41
6. A, terminal, day 2						

also note that there is a higher partial correlation between frequency of assistance on day 2 and errors on day 1 than between errors on day 2 and errors on day 1. This might possibly indicate that the amount of assistance was affected by the subjects' previous performance as well as their actual performance.

Discussion

The motive for this study was the idea that one important obstacle when performing database search pertains to the fact that the user may neglect, or even be unable, to analyse the queries properly. Subjects were given queries in written Swedish in the present study. Their task on day 1 was to translate them into one of the two different query languages. In real-life situations the user's task may be less well-defined than translating a query presented in written language into a formal query language. It is more likely that an information-seeking task is part of a larger task (e.g. writing a memo, decision making), and to formulate the information need explicitly in natural language may be an intermediate step before constructing a query in a formal query language. However, the necessity of making a logical analysis of the query remains. Studies related to the learning of database query languages, as a rule, have addressed the problem of translating queries in written language into a formal query language (e.g. Reisner 1977, 1981; Katzeff (1989). An obvious reason for this is the presence of many factors inherent in a naturalistic learning situation, which could make a controlled investigation not feasible.

The most pertinent result of the present study is that the effect of giving training in analysing queries in terms of deep-structure turned out to be dependent on type of query language being taught. Subjects trained on the graphic query language were found to benefit from pretraining related to 'deep-structure' in contrast to subjects trained on the textual query language. The usefulness of giving specific training in analysing queries may be related to the constraints imposed by the semantics and syntax of the query language with respect to the necessary way of analysing queries. Queries in the textual query language had to be constructed according to the following template:

GET_WHERE_EQUALS_(AND/OR/NOT_EQUALS_).

The keywords in this template more or less enforce a user to analyse queries in terms of an 'information-to-be-retreived and condition' scheme. The meaning of GET, WHERE, EQUALS in this query language seems to be congruent with the meaning of these words in natural language. The syntactical rules of the graphic query language (See fig. 1), on the other hand, must have appeared arbitrary for the subjects. There were no keywords, which conveyed information about what logical constituents to look for in the written query and about where to insert these constituents in the graph shown in fig. 1. This property of the graphic query language may have rendered the specific pretraining in analysing queries particularly useful.

It is noteworthy that some beneficial effect of pretraining on analysing queries under the condition 'graphic query language' seems to remain, though it is non-significant ($p = 0.053$) while the subjects wrote the program code on paper. It is likewise noteworthy that the impact of pretraining on 'deep-structure' appeared to be strongest at the onset of the query construction task, and then quickly decreased. The same tendency has been observed in two previous experiments, in which the effect of prior knowledge of informational content on performance of a database search task was studied (Linde and Bergström 1988, 1990).

It is also worth mentioning that Adelson and Soloway (1985) have investigated the role of task knowledge in software design. They gave programmers three different programming tasks. Task knowledge was defined as familiarity with the objects (e.g. a mail system) and/or familiarity with the domain (e.g. a library). The results indicated that programmers who lacked task experience had to perform effortful mental simulations of program functions in contrast to programmers with task experience, who were able to visualize the flow of information in the program in a more automatic fashion. These results are similar to the ones obtained by Linde and Bergström (1988). They also bear a relationship to the results obtained in the present study. Subjects given training on the graphic query language without pretraining in analysing queries might have had difficulties in establishing a useful representation of the task.

Appendix

Example of a query given to the subjects on day 1

In what activity was Anthony involved at 230 p.m.?
Formal query expression:

```
GET EVENTS.ACTIVITY WHE EVENTS.NAME EQ ANTHONY AND
EVENTS.TIME EQ 230 P.M;
```

Tasks given to the subjects on day 2

(1) Write a procedure, which always answers the query 'What is Anthony's profession?'
Program code:
```
GET PERSONS.PROFESSION WHE PERSONS.NAME EQ ANTHONY;
```

(2) Write a procedure, by which a user can pose a query about any person's profession.
Program code:
```
+ LET &A = 'NAME OF THE PERSON':
GET PERSONS.PROFESSION WHE PERSONS.NAME EQ &A;
```
VDU information (from the user's perspective):
NAME OF THE PERSON:

(3) Write a procedure, by which a user can pose a query about any column value for
any optional person in the table EVENTS.
Program code:
```
+ LET &A = 'NAME OF THE PERSON':
+ LET &B = 'OPTIONAL COLUMN':
GET EVENTS.&B WHE EVENTS.NAME EQ &A;
```
VDU information
NAME OF THE PERSON:
OPTIONAL COLUMN:

(4) Insert the procedures in a menu.
Program code of the menu:
```
+ CASE PROC1 'THE PROFESSION OF ANTHONY';
+ CASE PROC2 'THE PROFESSION OF AN OPTIONAL PERSON';
+ CASE PROC3 'OPTIONAL COLUMN OF AN OPTIONAL PERSON';
```
VDU information
(1) THE PROFESSION OF ANTHONY;
(2) THE PROFESSION OF AN OPTIONAL PERSON;
(3) OPTIONAL COLUMN OF AN OPTIONAL PERSON.

Classification of errors: Query construction

(1) DEL: missing or incorrect delimiter, (2) KOL: incorrect or missing column label, (3) LOG: incorrect logical operator, (4) KEY: incorrect or missing keyword, (5) INT: unnecessary interruption of a query, (6) PAR: missing or incorrect parameter, (7) REL: missing or incorrect relational operator, (8) MIS: conditional part or information-to-be-retrieved part completely missing, (9) SYN: multiple syntactical error, (10) TAB: missing or incorrect table label, (11) CON: one, of two conditions, missing or incorrect, (12) VAL: missing or incorrect value of the condition, (13) IRC: irrelevant condition, (14) IRP: irrelevant parameter.

Classification of errors: Procedure construction

ERRORS WHILE WRITING PROCEDURE CODE ON PAPER

(1) Syntax error due to improper use of query language, (2) incorrect definition of variable in variable definition line, (3) incorrect definition of variable in query line, (4) incorrect or misleading prompt, (5) others

ERRORS WHILE ENTERING PROCEDURE CODE TO THE SYSTEM

(6) incorrect labeling of the procedure, (7) program code of a 'new' procedure entered under the label of a previous procedure, (8) several, distinct lines of program code entered to the system as one line, (9) failure to enter proper mode, (10) mismatch between initiated transaction and mode state, (11) incorrect use of (line-oriented) editor, (12) other, formal syntax errors

ERRORS WHILE CONSTRUCTING MENU

(13) formal syntax error in +CASE line, (14) relevant procedure label omitted in +CASE line, (15) failure to understand why each procedure must be linked to a special +CASE line, (16) misleading or incorrect guiding menu text at a certain option, (17) mode error, (18) others

ERRORS WHILE TESTING PROCEDURES

(19) failure to understand one's own prompt text, (20) failure to understand one's own menu text, (21) inability to correct prompt text in a way that makes it comprehensible for the potential use, (22) inability to correct menu text in a way that it comprehensible for the potential user, (23) mode error, (24) incorrect use of editor, (25) others.

References

Adelson, B., 1981. Problem solving and the development of abstract categories in programming languages. Memory and Cognition 9, 422–433.

Adelson, B. and E. Soloway, 1985. The role of domain experience in software design. IEEE Transactions an Software Engineering SE-11, 1351–1360.

Craik, F.I.M. and R.S. Lockhart, 1972. Levels of processing: A framework for memory research. Journal of Verbal Learning and Verbal Behavior 11, 671–684.

Katzeff, C., 1989. Cognitive aspects of human computer interaction: Mental models in database query writing. Doctoral dissertion, Akademitryck, Stockholm.

Linde, L. and M. Bergström, 1988. Impact of prior knowledge of informational content and organization on learning search principles in a database. Contemporary Educational Psychology 13, 90–101.

Linde, L. and M. Bergström, 1990. 'An experimental study on mental models in database search'. In: D. Ackermann and M.J. Tauber (eds.), Mental models and human–computer interaction I. Amsterdam: North-Holland.

Reisner, P., 1977. Use of psychological experimentation as an aid to development of a query language. IEEE Transactions on Software Engineering SE-3, 218–229.

Reisner, P., 1981. Human factors aspects of database query languages. Computing Surveys 13, 13–31.

Rist, A.S., 1989. Schema creation in programming. Cognitive Science, 13, 389–414.

Acta Psychologica 78 (1991) 243–256
North-Holland

Adapting systems to differences between individuals *

Frances Jennings and David Benyon

Open University, Milton Keynes, UK

Dianne Murray

University of Surrey, Guildford, UK

Adaptive systems should be able to accommodate the preferred interface styles of different users. An experiment was conducted in order to determine whether significant differences exist between individuals performing the same task, using different interfaces. Individual users' performances on five different interfaces to a computer database system, after the initial learning stage, were compared with their scores on various cognitive and personality tests. The results suggested that two interface styles are necessary for database systems in order for them to suit a range of users: an aided-navigation interface with a constrained dialogue for low spatial ability users, and a non-aided navigation interface with a flexible dialogue for high spatial ability users. Both interfaces should minimize the amount of verbal input necessary.

The work described here is part of a project concerned with developing a generic architecture for adaptive systems; computer systems which automatically adapt aspects of their functionality and/or interface to suit the characteristics of individual or groups of users. This architecture is reflected in a software support system for the adaptive

---

* This work forms part of a project entitled Adaptive Systems and User Modelling Tools, which is funded by the National Physical Laboratory (NPL) under EMRA 82-0486.

This work was completed as part of Jennings' thesis (Jennings 1991). Although she is careful to present the results in the context of the user group (all graduates) and the difficulty of the queries, the results demonstrated that a non-navigational, constrained interface was more suitable for the group of users with a low spatial ability score who also had low command experience and were occasional users. The SQL type interface was suitable for users with a high spatial ability, for users with a low spatial ability and high experience of command languages and for those with a low experience, but regular computer use (since this will soon increase their experience).

Requests for reprints should be sent to D. Benyon, PACIS Research Group, Computing Dept., Open University, Milton Keynes MK7 6AA, UK.

system developer known as an *adaptive system development environment* (ASDE, Benyon and Murray 1991; Benyon et al. 1990).

Our architecture of an ASDE (and of adaptive systems in general) consists of three main components. The *domain model* describes the logical and physical features of the application such as the functions which the system can perform and the objects which may be manipulated. The *user model* represents appropriate characteristics of the system users. The *interaction model* governs the relationship between users and the domain and contains three types of rules; *inference* rules which infer characteristics form the user-system interaction, *adaptation* rules which adapt the interaction to suit different users and *evaluation* rules which evaluate the suitability of actual or possible adaptations.

We have shown previously (Benyon et al. 1988) that the mechanisms of adaptive systems are relatively straightforward. The difficulties with such systems lie in establishing reliable and realistic relationships between user characteristics and system features. Once these have been identified adaptation rules of the form

IF⟨user has characteristic X⟩THEN⟨system has feature Y⟩

can be formulated.

The purpose of this paper is not to consider the mechanisms of adaptivity, nor the theoretical background of adaptive systems. A comprehensive discussion of adaptive systems is in preparation (Benyon and Murray 1992). In this paper, we concentrate on an examination of the effects that individual and measurable differences between users have on the interaction with the system. Given that such differences can be found, we can then turn our attention to finding methods of eliciting these characteristics from the users. This can take the form of asking the user, users supplying their own user model on a device such as a smart card, or by the system inferring characteristics directly from the interaction (which may subsequently be confirmed explicitly with the user). The feasibility of this latter approach has been suggested by Vicente and Williges (1988) who found that the use of particular commands correlated significantly with a user's level of spatial ability. If such a relationship can be confirmed then suitable inference rules can be established.

In order to obtain the information necessary to design alternative interfaces for a computer database system to suit a range of users, an exploratory experiment was conducted. Twenty four university research students and research workers, aged between 25 and 40, (sixteen male and eight female) used a database system which had five different styles of interface. The subjects were all regular computer users, familiar with both mouse and keyboard input. This subject group was selected as the study was not concerned with users' learning of computer systems. The subjects needed to be capable of learning a system quickly to a reasonable degree of proficiency, and had to be able to type and use a mouse.

The task

Database systems are a class of systems which are generally used by a wide variety of people for a wide variety of purposes. It therefore makes sense to try to adapt these systems to the people who use them.

The database system developed for this experiment supported the single task of obtaining lists of items available from a mail-order shopping catalogue. Uses had to specify the type of item they were interested in and values for three of its attributes. The users were required to query the database and obtain information such as:

- How many types of women's t-shirts are available, which cost less than £15, are navy in colour and are U.K. size 14-16? or
- Are there any carpets available, which cost less than £10 per yard, are brown in colour and 13 feet in width?

The interfaces

The database system had five different styles of interface, designed to represent a range of the typical styles of interfaces to current computer systems. The five interfaces developed for this system are presented in table 1. Each of the interfaces was evaluated by an independent HCI expert who confirmed that they were all well-designed and typical of the interface style.

Having five representative interfaces provided a number of benefits. Firstly, user characteristics could be looked for which affected users' use of many interface styles, not just one. Secondly, it allowed perfor-

Table 1
Summary of the five interfaces.

Interface style	Examples of this style	Characteristics
Button interface	Macintosh Hypercard applications	Users select an item type by clicking the mouse button on named boxes representing narrowing choices of categories of item types, and clicking on their required values for the attributes of the item type.
Command interface	MS-DOS, Unix etc.	Users type in a syntactically correct statement specifying the type of item they are interested in and the values for its three attributes.
Iconic interface	MacPaint and MacDraw on the Apple Macintosh	Users click on pictures of the item types they are interested in, and then select attribute values as for the mouse and button interface.
Menu interface	Many Apple Macintosh applications	Users select an item type from a 'walk-through' menu of category choices, and pull-down menus for selecting attribute values.
Question interface	Many commercial data entry systems	Users answer a series of questions about which category, item type and attribute value choices they require.

mance differences due to the task to be distinguished from performance differences due to the interface style. All the interfaces supported the same underlying task, so if a particular user characteristic had the same effect on performance for all the interfaces, this would suggest that it was the task causing the effect, whereas if the effect on performance was different for the different interface styles, this would suggest that it was indeed due to the interface style.

Thirdly, the interfaces on which users with a particular characteristics did well could be compared with the interfaces on which they did not do quite so well. The aspect or aspects of the interface style causing the effect would be likely to be aspects which were similar in the interfaces the user performed well on, but different from the interfaces on which they performed less well on. The interfaces could therefore be analysed in terms of the aspects in which they were similar to and different from the others.

Fourthly, a comparison of the aspects of the interfaces with which the users with particular characteristics had difficulties was made with the other interfaces which the users performed better on. This immediately suggested an alternative way of presenting the aspect.

User characteristics

Individual differences appear to have a big impact on human–computer interaction (Egan 1988). It has been suggested (Van der Veer 1990) that some characteristics of people are more resistant to change than others. In particular, cognitive and personality traits are most resistant to change. It makes sense to adapt systems to these characteristics since they are most difficult for the user to alter. Five cognitive and personality characteristics were selected from a review of previous research as being the most likely to have relevance to users' use of different interface styles and were tested in this experiment.

Spatial ability is a cognitive characteristic which offers a measure of a user's ability to conceptualize the spatial relationships between objects. It is closely allied to the notion of a cognitive map (Neisser 1976), and (following Vicente et al. 1987; and Vicente and Williges 1988) may also be related to a user's ability to navigate through a complex space. Vicente et al. (1987) also found that verbal ability (a measure of a user's linguistic skills) correlated significantly with a user's performance on their experimental task of searching for an item in a hierarchical file structure. Field dependence measures the ability of an individual to separate an item from the background. Fowler and Murray (1988) suggested that this characteristic may influence a user's preferences for a style of interface. Short-term memory capacity measures the number of chunks of information which an individual can hold. Benyon et al. (1987) found that a user's short-term memory capacity may influence his or her ability to digest data displayed. Logical-intuitive cognitive style may also effect the preferred way of viewing data. Garceau et al. (1988) found that subjects with a logical cognitive style were able to locate a particular item with a tabular representation of data than with a graphical representation.

Method

Each subject was given all five interfaces to use. Before the subject was presented with the task, the purpose and the procedure of the experiment were explained to the subject. The five interfaces were presented to each subject in a random order. For each interface the subject was given a practice session followed by a test session. The practice session consisted of the subject being given a series of tasks to carry out using the computer system which were comparable to the tasks they would be asked to carry out in the test session. The experimenter was available to provide any necessary help and to answer any questions the subject had. The test session started when the subject was confident that they had learned how to use the interface. For the test session the subject was given a series of tasks to carry out on their own using the system. During the test session the experimenter sat away from the subject behind a screen. The subjects were given different but equivalent test tasks for each interface, to minimize practice effects while maintaining comparability. The test tasks (all of the same form as that described earlier) were balanced in terms of the type of tasks, the number of steps required to complete the tasks, the amount of typing involved and the catalogue categories the tasks concerned. The complete session for all five interfaces lasted approximately one hour.

The subjects were then given psychological tests (NFER-NELSON, Myers-Briggs, Saville and Holdsworth) to determine their positions on the selected cognitive and personality variables. In addition, the subjects were given a questionnaire asking them to rate their experience of using interfaces similar to each of the five types of test system interfaces. Experience could be rated as none, some or a lot.

The subjects' success at performing the computer tasks was assessed by the time taken to complete the tasks (the measure found to be most useful by Vicente et al. 1987; Vicente and Williges 1988). This was taken to indicate the level of difficulty they experienced; the assumption being that the higher the mental load the longer the subject would take in determining the appropriate input and the more errors they would make. The time differences could be used to indicate the *comparative* difficulties subjects had within single interfaces. However time differences could not be looked at to examine the difficulties between interfaces, as each interface took a different amount of time to carry out tasks due to its actual design. For example the menu interface required more mouse clicks than the iconic interface to carry out the same task.

As well as this quantitative measure, the subjects were also given a questionnaire asking them to rate how easy and how enjoyable they found each of the interfaces to use on scales of one (most easy/enjoyable) to five (least easy/enjoyable).

Results

A number of analyses of the data were carried out.

Firstly, Pearson product moment correlation coefficients were calculated between the subjects' scores on the psychological tests and the times they took to complete the test tasks with each of the interfaces. A significant correlation here would suggest a link

Table 2
Correlations between test scores and task completion times for the five interfaces.

	Interface				
	Button	Iconic	Menu	Question	Command
Spatial ability	−0.37 [a]	−0.02	−0.07	−0.48 [b]	−0.58 [c]
Verbal ability	−0.07	0.06	−0.13	−0.43 [b]	0.01
Field independence	−0.19	0.20	−0.07	−0.08	−0.39 [a]
Short-term memory	−0.08	−0.29	−0.34	−0.15	0.03
Thinking/feeling	0.08	−0.36 [a]	−0.13	−0.08	−0.08

[a] $p < 0.1$;   [b] $p < 0.05$;   [c] $p < 0.01$.

between a characteristic and subjects' performance on an interface. The results, presented in table 2, show a particularly strong relationship between spatial ability and use of the command language interface.

Secondly, correlation coefficients were calculated between users' self-reported levels of previous experience with similar types of interface in order to determine whether or not previous experience affected subjects' performance on the interfaces. Although no significant correlations were evident from the data, suggesting that previous experience does not affect performance after the learning process is complete, this result was not totally confirmed in a subsequent experiment (Jennings 1991).

Thirdly, the subjects' scores on the psychological tests and their ratings of ease and enjoyment for the interfaces were analysed in order to see if there were any links suggested between the users' cognitive and personality characteristics and how they felt about the interfaces. Only one significant correlation was found; between scores on the field independence test and the self-rated ease of the question interface ($r = -0.53$, $df = 22$, $p < 0.05$). Subjects with a high field independence rated the question interface as easier than those with a low field independence. However, a sample t-test (two tailed) between the mean ease ratings for the question interface for the low and high field dependency groups did not show a significant difference ($t = 1.39$, $df = 22$, $p > 0.1$).

Fourthly, a correlation coefficient was calculated between the users' ease and enjoyment ratings of the interfaces and their task completion times for the interfaces, to see if how the users felt about the interfaces correlated with how well they actually performed on the interfaces. This analysis showed that subjects' ratings of how they felt about the interfaces did not correspond well with how they actually did on the interfaces.

Finally, the subject group was divided into two according to their scores on each psychological test – the twelve highest scorers in a *high* group and the twelve lowest scorers in a *low* group for each characteristic. The mean task completion times were then calculated for the high and low groups on each characteristic for each interface, and two sample t-tests were used to test for significant differences in the mean times for the high and low groups on characteristics for interfaces which the correlations had suggested the characteristics may affect performance on. Both of these analyses were used as, even though the correlations alone could be used to suggest that a characteristic does affect subjects' performance on a particular interface, this result is only useful if users who are high and low on this characteristic have a large difference in the mean times they take to complete the tasks using the interfaces. There is no point in adapting an interface to accommodate a user characteristic which does not in fact produce much of a difference in users' performances. Graphs of these analyses are presented in fig. 1.

Discussion

From table 2, it can be seen that scores on the spatial test correlated significantly with performance on both the question and the command interfaces, and approached a significant correlation with performance on the button interface. In each case the higher score relating to faster performance. Scores on the verbal test correlated significantly with performance on the question interface – a high verbal score relating to faster performance. Scores on the thinking/feeling scale approached a significant correlation with performance on the iconic interface, people towards the thinking end of the scale performing faster.

The results presented in fig. 1 show that the high and low groups on the spatial ability test differed significantly in their performance on the command interface ($t = 3.34$, $df = 22$, $p < 0.01$), and approached a significant difference in their performance on the question ($t = 1.88$, $df = 22$, $p < 0.1$) and button ($t = 1.73$, $df = 22$, $p < 0.1$) interfaces. The high and low groups on the verbal ability test were found to differ significantly in their performance on the question interface ($t = 2.11$, $df = 22$, $p < 0.05$). However, the high and low groups on the field independence test were found not to differ significantly in their performance on the command interface ($t = 1.20$, $df = 22$, $p > 0.1$), despite the correlation between the field independence score and command interface task time approaching significance. The high and low groups on the thinking/feeling scale (high = thinking, low = feeling) were also found not to differ significantly in their performance on the iconic interface ($t = 1.60$, $df = 22$, $p > 0.1$).

The results show that spatial ability and verbal ability significantly affected subjects' performance when using the computer systems, and suggest that field dependency may affect performance, there being a difference of 41 seconds between the mean task completion times for the low and high field independence groups on the command interfaces, although the difference did not reach statistical significance. Thinking/feeling and short-term memory capacity (as measured here) were found not to affect subjects' performance on the interfaces. The mean task completion times for the low thinking and high thinking groups only differed by 28 seconds which was not significant, although the correlation between position on the thinking/feeling scale and time taken on the iconic interface did approach significance.

The effect of presentation style found by Garceau et al. (1988) did not therefore transfer to this experiment. In fact the direction of the difference in the means in this experiment was opposite to that which would have been expected from Garceau et al.'s results. The spatial ability, verbal ability and field independence scores did produce different effects for different interfaces, suggesting that it was the *interface style* which

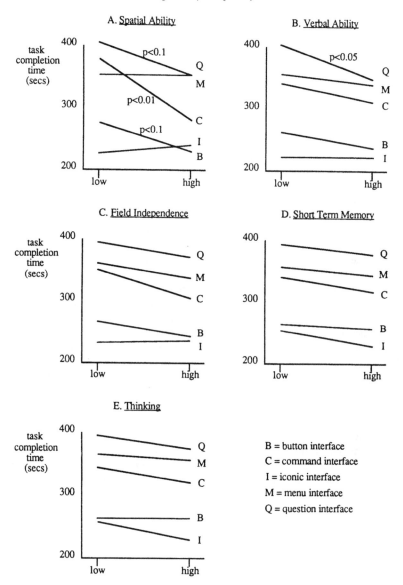

Fig. 1. Graphs of mean task completion times (in seconds) for the high and low scoring groups on each test for each interface.

was causing the effects, and not the underlying task (since the tasks were equivalent in each case).

From fig. 1, graph A, it can be seen that people with high spatial ability do better on the command, question and button interfaces than people with low spatial ability, but do equally well on the iconic and menu interfaces. The command, question and button interfaces all require significant *navigation* through the system to achieve the tasks, whereas the iconic and menu interfaces require this to a much lesser extent. The command language interface has three levels, or modes; the system mode from which the catalogue can be accessed and to which the user returns when they quit from the catalogue, the catalogue mode where the user can specify the catalogue items they are interested in, from where they can access the catalogue help system, and to which they return when they quit from the help system. The help system constitutes the third mode. Although different system prompts indicate the different modes, the user has to keep in mind the mode they are at, and how the modes link together. For example, a user cannot go straight from the system mode to the help mode without going through the catalogue mode.

The button and question interfaces require the user to go through a series of hierarchical category choices in order to reach the item type they are interested in. The user has to understand this structure and be able to navigate around within it. The iconic interface however, after an initial choice of two categories, presents a single visual scene from which the user can pick out the type of item they are interested in. The menu interface involves a walk-through menu in order to select the desired item type. The hierarchy of category choices is displayed as the user walks through the menu so that the structure does not have to be clearly remembered by the user.

These results suggest that spatial ability could relate to a users ability to cope with an interface requiring navigation. Being able to cope with such navigation involves knowing where things are in the structure and how to move through the structure efficiently to reach them. This is similar to the suggestion by Van der Veer (1990) of a relationship between spatial ability and mental representation of complex systems.

The difference in performance between the high and low spatial ability groups was much greater for the command interface than for the button and question interfaces. This suggests that there is something additional about the command interface which is causing difficulty for users who have a low spatial ability. The command language interface involves a much less *constrained* dialogue than the button, question, menu and iconic interfaces. In these, the interaction is quite structured. With the button, question, menu and iconic interfaces the user is presented with screens clearly indicating whether a category, item type or attribute choice should be made next, or an operation used to open, close or return to the start of the catalogue. However, with the command interface a system prompt is displayed at all points, with little indication of what type of input is expected next from the user.

Spatial ability could therefore also relate to a user's ability to cope with interfaces allowing a very open and flexible dialogue. Being able to cope with a flexible dialogue involves being clear when and where particular input can be used to best advantage and what its outcome will be. However, this latter result may be due to the fact that the users' spatial ability and field independence scores correlated significantly. (See table 3 for Pearson product moment correlation coefficients calculated between subjects' scores on the psychological tests.) Eighteen out of the 24 subjects were in the same group for spatial ability as they were for field independence i.e. in the high or low groups for both characteristics. It may in fact be field dependency which relates to the ability to cope with a flexible dialogue, as suggested by Fowler and Murray (1988).

Table 3
Correlations between subjects' scores on the psychological tests.

	Verbal ability	Field independence	Short term memory	Thinking
Spatial ability	0.22	0.53	0.18	0.02
Verbal ability		−0.05	−0.08	0.35
Field independence			0.27	0.07
Short-term memory				−0.16

Graph C, fig. 1, lends some support to this idea. Subjects with a high field independence performed faster on the command interface than subjects with a low field independence, although the result did not reach significance, while performing similarly on the other interfaces. The alternative argument is that these effects for the field dependency scores are due to the correlation between spatial ability and field independence. However, if this was the case, a correlation of some degree would be expected between the field dependency scores and performance on the button and question interfaces as well. The correlation between field dependency and performance on the question interface is so low ($r = -0.08$, $df = 22$), that it seems likely that field dependency is indeed causing a separate performance effect for the subjects on the interfaces from spatial ability. The question interface, although hierarchical, has a very constrained dialogue.

From fig. 1, graph B, it can be seen that people with a high verbal ability do better on the question interface than people with a low verbal ability, but do equally well on the button, menu, iconic and command interfaces. The question interface involves the entering of a large amount of accurate verbal information which is not involved in the button, menu and iconic interfaces which are mouse operated, or in the command interface where a short statement specifies the type of item a user is interested in. This suggests that verbal ability could relate to a user's ability to cope with interfaces requiring a large amount of accurate verbal input.

These results suggest that aspects such as the amount of navigation required by an interface, the flexibility of the dialogue and the amount of verbal information to be entered could account for the observed individual differences in performance.

Conclusion

From the performances of the high and low spatial ability groups on the five interfaces, it can be seen that any of the five interfaces would be suitable for the high spatial ability group who are mostly high on field independence as well, but that users with a low spatial ability who are mostly low on field independence should avoid interfaces such as the button, question and command interfaces which contain a navigational component and also a flexible dialogue structure in the case of the command interface. For the high verbal ability subjects, again any of the interfaces are suitable, but low verbal ability subjects should avoid interfaces such as the question interface which involve a large amount of verbal input.

These results suggest that a non-navigational, inflexible dialogue without a large amount of verbal input would be suitable for all users. The iconic interface in this experiment, on which all subjects performed equally well, fits these requirements. This interface was also quicker for completing the tasks than the other interfaces. If one interface can be developed which is suitable for all users then why not just present this to users as a single fixed interface and forget about having an adaptive system?

The problem with this argument is that the data in a database usually requires a representation which involves the linking of several files and therefore requires a degree of navigation in order to elicit information from the database. The iconic interface in this experiment only avoided this navigational component as it was attached to an unrealistically small and structurally simple database (one file with four fields). In most databases it would not be possible to produce a system which contained no navigational component. In such cases an aid for navigating a system would have to be given to users low on spatial ability, which would not be necessary for users high on this characteristic. Also, although high spatial ability users performed well with the highly constrained interfaces, if a large amount of work was being done with the system, it would prove very laborious for users who have the ability to take short cuts to go through set routines. A second point to emerge from this experiment is that it seems sensible to reduce verbal input in all cases, either by allowing the use of abbreviations or by using function keys.

The results of this experiment suggest that two forms of interface are needed for a database system in order to accommodate the variety of users. A navigation-aided system with a fairly constrained dialogue is required for users with a low spatial ability who are mostly field dependent as well, and a non-aided navigational system with a flexible dialogue structure for users with a high spatial ability who are mostly field independent. This conclusion fits in with that of Vicente et al. (1987), who suggested a link between spatial ability and the ability to navigate a hierarchical structure and with the work of Fowler and

Murray (1988) who suggested a link between field dependency and the ability to cope with an open and flexible dialogue.

The experiment found that previous experience was not important in determining the best interface style for a user. This result is also supported by the findings of Vicente et al. (1987), who found that previous experience had no effect on their file searching task when cognitive characteristics were partialled out. It is likely that previous experience has an effect during the initial learning of a system but not thereafter. However, gaining experience with a particular interface style may be able to compensate for the cognitive demands which an interface makes of its users (Benyon 1990).

This experiment was designed to generate ideas as to how interface styles relate to user characteristics. Many of the effects found in this experiment, apart from that for spatial ability and performance on the command interface, were not statistically very strong. Due to the number of post hoc comparisons carried out, there is some probability that the results of the experiment contain one or more erroneous conclusions and show to some extent the limitations of the statistical techniques employed. The results, therefore, do not prove any definite relationships between interface style and cognitive characteristics, but they do suggest the sort of relationships which may exist.

Further work needs to be done to test the suppositions from this experiment, to see if they are indeed valid. This can be done by constructing two interfaces to a realistically sized database system, which fit the outlines given above, and seeing whether users with high and low spatial ability do perform better on the interfaces designed for them. The standard database language SQL offers a flexible dialogue which requires navigation and a form-filling or question and answer dialogue would provide the non-navigational, inflexible dialogue.

References

Benyon, D.R., 1990. 'Adapting systems to differences between individuals'. In: Proceedings of the Australian Society for Cognitive Science conference, University of NSW, Sydney, Australia, November.

Benyon. D.R. and D.M. Murray, 1991. 'An adaptive system development environment'. In: R. Damper (ed.), Software tools for interface design. New York: IEEE Press.

Benyon, D.R. and D.M. Murray, 1992. Adaptive systems. International Journal of Man–Machine Studies (in press).

Benyon, D.R., P.R. Innocent and D.M. Murray, 1988, 'System adaptivity and the modelling of stereotypes'. In: H.J. Bullinger and B. Shackel (eds.), Human–computer interaction – Interact '87. Amsterdam. North-Holland.

Benyon, D., S. Milan and D. Murray, 1987. Modelling users' cognitive abilities in an adaptive system. Paper presented at the fifth Empirical Foundations of Software Sciences (EFISS) conference, Risø National Laboratory, Denmark.

Benyon, D.R., D.M. Murray and F. Jennings, 1990. 'An adaptive system developer's tool-kit'. In: D. Diapen (ed.), Proceedings of Interact '90. Amsterdam: North-Holland.

Egan, D.E., 1988. 'Individual' differences in human–computer interaction'. In: M. Helander. The handbook of human–computer interaction. Amsterdam: North-Holland.

Fowler, C.J.H. and D. Murray, 1988. 'Gender and cognitive style differences at the human–computer interface'. In: H.J. Bullinger and B. Shackel (eds.), Human–computer interaction – Interact '87. Amsterdam: North-Holland.

Garceau, S., M. Oral and R.J. Rahn, 1988. The influence of data presentation mode on strategic decision making performance. Computer Operations Research 15, 479–488.

Jennings, F., 1991. Database systems. Different interfaces for different users. M. Phil Thesis, Open University, Milton Keynes (unpublished).

Murray, D. and D. Benyon, 1988. 'Models and designers tools for adaptive systems'. In: Proceedings of ECCE4, Cambridge, UK, September 1988.

Myers-Briggs Type Indicator. Oxford: Oxford Psychologists Press.

Neisser, U., 1976. Cognition and reality. San Francisco, CA: W.H. Freeman.

NFER-NELSON. Abstract ability test. ASE Division.

NFER-NELSON. Verbal ability test. ASE Division.

Saville and Holdsworth Ltd. Spatial ability test ST7.

Van der Veer, G.C., 1990. Human–computer interaction. Learning, individual differences and design recommendations. Alblasserdam: Offsetdrukkerij Haveka.

Vicente, K.J. and R.C. Williges, 1988. Accommodating individual differences in searching a hierarchical file system. International Journal of Man–Machine Studies 29, 647–668.

Vicente, K.J., B.C. Hayes and R.C. Williges, 1987. Assaying and isolating individual differences in searching a hierarchical file system. Human Factors 29, 349–359.

Text editing

Acta Psychologica 78 (1991) 259–285
North-Holland

Interference among text-editing commands: Fan-effects and the role of system consistency

Martin Heydemann, Rudolf Hoffmann and Rainer Schmidt *

Technische Hochschule Darmstadt, Darmstadt, Germany

Interference effects among several commands offered by an experimental text editor were demonstrated. The experiment involved 24 computer novices correcting texts with a computer in several working sessions. The interference effects are interpreted as 'fan-effects', similar to those studied with propositional material (Anderson 1976). During the processing of editing tasks, increased latencies for key-strokes were observed with rising degrees of fan. Two sorts of fan degrees were manipulated in the experiment: Firstly, fan-effects arise if the same features of editing tasks occur for different commands (*task fan*). Secondly, fan-effects also arise if the same user action (e.g. key-stroke) is required for the execution of different commands (*key fan*).

The mapping from task features to command names was varied. Consistent editors were compared to inconsistent ones. In consistent editors the same task features were mapped to the same key-strokes for different commands, while in inconsistent editors the mapping from task features to command keys was random. For consistent editors, interference due to fan is reduced as compared to inconsistent editors. The interference effects, as well as their qualitative dependence on consistency, point to functional equivalences among interference effects as observed here and as studied previously with propositional materials.

Users of word-processing software are required to know a large number of commands. Several different types of command interfaces are available in computer systems: direct manipulation, menu-based systems and command language systems (Shneiderman 1987). Menu-based systems are preferred by novices or occasional users, however, an experienced user can work more quickly with a command language, once the commands have been mastered (see Streitz et al. 1987). The necessary proficiency can only be achieved if the user has memorised all commands and requires no additional reference.

* We are grateful to the Deutsche Forschungsgemeinschaft who provided the funds (Schm 350 2/3) which made this work possible.

Requests for reprints should be sent to M. Heydemann, Fachbereich 3, Technische Hochschule Darmstadt, Steubenplatz 12, D-6100 Darmstadt, Germany.

This paper investigates the performance of users in operating command language-based software. The investigation was carried out using text editors running on personal computers. The subjects are to have learned the commands sufficiently, such that they can retrieve them quickly from memory with a minimum of error. The central topic of this research is to determine to what extent memory recall processes affect the ability to use command languages for tasks such as text editing.

Positive and negative transfer for recall processes

With memory recall research the question arises whether there are mutual influences between different memory traces. These influences could result in either positive transfer or, conversely in interference. In classic psychological research both positive and negative transfer are evident (e.g. the transfer surface of Osgood 1949). Interference was thoroughly investigated in A–B to A–C-transfer paradigms with paired-associate learning (for an overview, see Crowder 1976: ch. 8). Here, subjects first learn an association between a stimulus and a response and then another association between the same stimulus and another response. Thus, two responses to the same stimulus are learned. If the responses are dissimilar, it is more difficult to recall the first response than in a condition in which only one response is learned. Interference also takes place during sentence learning, when the same concepts (e.g. the same subject, verb, or object etc.) occur in different sentences, and subjects are to recognise these sentences. This interference is called fan-effect (e.g. Anderson 1976: ch. 8; 1983). The learning of text editing exhibits similarities to the learning of paired associates: if a specific stimulus situation is presented, e.g. a particular editing task, then the user has to respond with the assigned key.

Positive and negative transfer in editing

Positive and negative transfer between user learned commands can also be expected in editors. Positive transfer means that learning and usage of commands has a positive effect on further command learning and usage. Interference is mutual negative influence between commands. Investigations showed predominantly positive transfer in editing. In general, initially learned commands require more practice than subsequently learned commands (Polson et al. 1986; Polson and Kieras 1985). Positive transfer also arises when commands are practised in the context of another system. People having prior editor experience learn a new editor more quickly than a control group with no prior editing experience (e.g. Karat et al. 1986; Polson et al. 1987; Singley and Anderson 1985; Ziegler et al. 1986). Prior knowledge may facilitate the learning of editing commands. For example, if command names are specifically related to descriptions of the operation that are compatible to the user's prior knowledge, then learning of command names is facilitated (Grudin and Barnard 1984). Also, letting users choose their own command names appears to help them remember the names (Jones and Landauer 1985).

Little negative transfer (i.e. interference) has been found, with that which does exist, appearing to be restricted to very specific instances. For example, Waern (1985) reports that experience in type-writing can lead to the incorrect use of the space bar instead of the cursor. Schmidt et al. (1991) show that processing time is prolonged when the user frequently switches between various editors. This effect only occurs directly after an editor switch and seems to disappear after a few instances of a given command. Landauer et al. (1983) report difficulties that arise when the same command names are used for operations that require different syntax. Singley and Anderson (1987–88) attempted to show negative transfer with an experiment in which users had to switch to a text-processing systems where all key assignments had been changed; nevertheless a large amount of positive transfer was found. In their study, 'negative transfer was restricted to the deletion components, where a nonoptimal method was imported from the training editor' (Singley and Anderson 1989: 134).

Theoretical considerations of transfer in text-editing

Despite the somewhat superficial similarities between paired-associate learning and command retrieval in editing mentioned above, there are marked differences between the two paradigms. Complex planning processes that include the setting of subgoals and the selection of methods are important in influencing the user's editing behaviour. Comprehensive models that aim to predict user performance take these factors into account (e.g. the GOMS-Model of Card et al. 1983). Referring to their experiment cited above, Singley and Anderson conclude, that when practising a skill such as editing, 'negative transfer should be largely restricted to a specific type of Einstellung where a particular method acquired during training turns out to be legal but nonoptimal in the transfer task' (Singley and Anderson 1989: 223).

A possible explanation for the dissimilar results between paired-associate/sentence learning and editor learning is provided by the distinction between declarative and procedural knowledge (Anderson 1983). Paired-associate/sentence learning is considered to be primarily declarative. This has the consequence that 'it is also clear that procedural knowledge behaves differently than declarative knowledge in that it does not seem to be subject to principles of interference' (Anderson 1987: 201). Regarding the differentiation between procedural and declarative knowledge, existing theories of text editing refer only to positive transfer (e.g. Cognitive Complexity Theory, Kieras and Polson 1985; Kieras 1988). Green and Payne's (1984; Payne 1989) Task Action Grammar (TAG) claims that an efficient mapping between command meaning and syntax that results in a minimum of syntax rules improves the learning of text editors. Again, interference is not considered in their theory.

Aim of the present investigation

It is a crucial assertion that a different type of knowledge is used in editing than is used in paired-associate/sentence learning. This is however based upon only a small number of experiments. This investigation will attempt to show interference occurring through memory recall in editing. It will then be carefully examined whether this interference is actually different than the interference occurring in declarative knowledge. The experiment focuses on editing skills as are actually required for editing on computers. Experiments with paper and pencil do not fulfil these requirements. Walker and Olson (1988) find a small amount of retroactive interference in an editing experiment where learning and testing of command names took place with paper and pencil. Their experiment more closely resembles paired-associate learning than computer editing. Thus the experiment of Walker and Olson seems to be more applicable to declarative rather than to procedural knowledge. It remains to be determined whether this interference occurs during real editing on computers.

Predictions based on research into fan-effects

We are interested in the skill of editing. Our study investigates editing at a stage, in which users are already able to utilize the editor correctly. That means that commands must be overlearned to a degree where failures occur due to occasional lapses rather than not knowing the correct operation sequences. In order to compare editing and declarative knowledge, a corresponding area of research in declarative knowledge must be found in which interference arises with overlearned material. Investigations of the fan effect fulfil this requirement. The fan effect was examined using sentences that in their simplest versions were equal to paired associates, e.g. Anderson (1974). In this experiment associations between names of people and locations were to be learned, e.g.:

'A hippie is in the park.
A hippie is in the church.
A policeman is in the park.'
   (Anderson 1974: 452)

In the experiment, subjects overlearned these sentences. Recognition tests were then performed. Subjects required extended recognition times when concepts (e.g. policeman, hippie, park) of the test-sentence also occurred in other sentences. This extended recognition time due to multiple occurrences of an element is referred to as fan-effect. The fan degree of a concept is equal to the number of propositions learned containing that concept. (For this example, one proposition represents the meaning of one sentence.) In the example above, 'park' has a fan degree of 2, and 'church' a fan degree of 1. In general, the fan degree of a concept may be defined to be equal to the number of memory traces containing the concept.

The fan effect asserts that raising the fan degree of the retrieval cues that are needed to identify a memory trace will result in extended processing times. Fan-effects with sentences occur not only in recognition tests but also in cued recall tasks. For instance, fan-effect can also be observed if a concept of a sentence is missing and has to be recalled (Anderson 1983: 116; Heydemann 1989).

Fan effects in declarative knowledge exhibit several characteristic properties. For a comparison between interference that may be found in editing and interference found in declarative knowledge, qualitative aspects should be taken into account. Three main results that will be referred to in this work are noted. *First*, with sentence materials, the largest increase in interference is found as the degree of fan increases from 1 to 2. Further rises in degree of fan yield relatively small rises in interference (e.g. Aschermann and Andres 1990). *Second*, fan-effect decreases with practice but does not appear to fade completely even after extensive training. Pirolli and Anderson (1985) showed fan-effects persisting even following 25 learning sessions. *Third*, fan-effects are prominent when completely unrelated sentences must be learned. Meaningful connections between sentences may reduce the amount of interference. For example, if sentences are part of a meaningful story, recognition latencies may be largely independent of fan degree (Myers et al. 1984; Smith et al. 1978). Also, if several sentences containing a certain concept have a common theme, and correct recognition depends only on identifying this theme, then the fan effect is reduced (Reder and Anderson 1980; Reder and Ross 1983).

The present investigation does not seek to reduce text editing to performance on sentence retrieval tasks. Nevertheless, if special consideration is given to the aspects of text editing and sentence retrieval tasks that are similar, interference may be expected to occur in text editing. There is one important structural similarity between sentence learning and learning of editing commands: Cues that are needed for information retrieval are associated with several responses and thus have varying degrees of fan.

Basic theoretical assumptions and hypotheses

Editing tasks are usually modifications to a text, e.g. 'insertion of words', 'copying of blocks' or 'deletion of lines' etc. The processing of an editing task can be divided into a sequence of individual user actions. At any given time the user is in an information state which includes the user's entire current knowledge. Each operation results in a change in this knowledge state. In principle, the user derives each operation step from the current knowledge state. This will be illustrated with an example. Assume that first a particular task, e.g. 'delete word', has to be recognised. Next, the cursor is to be positioned on the relevant word. The user will be able to do this as soon as the task has been recognised and the word identified on the screen. For the second step, the user needs the additional information that the cursor has been correctly placed in order to press the appropriate keys. Thus, the basic principle is simple. Each user operation is deduced from the currently available information contained in the user's present knowledge state; the more difficult the deduction, the higher the strain on the user.

One factor that affects the evaluation of the information state is the degree of 'fan'. Fan refers to the case when the same feature or single information element of different information states is connected to different operations. In other words, a single information feature is connected with competing user responses. In this example, the user may learn the key sequences 'Ctrl D' for 'delete word' and 'Ctrl T' for 'delete line'. Here the task feature 'delete' is connected with two competing actions, pressing key 'D' or key 'T'. This should cause interference as compared to an editor that only contains one deletion command. Generally, interference should arise when the fan of information elements connected to a user operation is larger than 1. As the degree of fan is increased, the amount of interference should be expected to rise. All features of the knowledge state that support the correct selection of an operation are information elements. If a user operation (e.g. a key-stroke) restricts the number of possible subsequent operations, then the applied operation serves as part of the information used for the selection of the next operation. Thus not only features of the editing task or the system feedback but also user operations themselves serve as information elements.

Two cases pertaining to different levels of analysis will be considered in the following text. Firstly, the fan of features that characterise the editing *task* may vary (*task fan*). Secondly, the elements of the user *operations* may have different fan degrees (*operation fan*).

Task fan (TF) contains all the information elements that are used for task representation. The fan degree of one task component depends on the number of different commands with the same component. Instances of task components are the types of task objects (e.g. 'word', 'line') as well as the functions that are applied to the objects (e.g. 'delete', 'move'). The number of commands with the same task features defines their degree of fan (see fig. 1a).

Operation fan includes all elements that are activated by the user during task processing. In the following we concentrate on the structure of key-strokes, which will be referred to as *key fan* (KF). The fan degree of one key is defined to be equal to the number of commands that contain that key (see fig. 1b). For instance, if the command

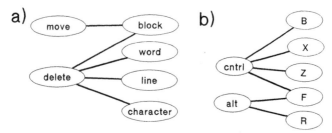

Fig. 1. Degrees of task fan (TF) for five typical commands (a). The links fanning out from a task element indicate the degree of task fan (TF = 4 for 'delete', TF = 2 for 'block', and TF = 1 for 'move', 'word', 'line', 'character'). Degree of key fan (KF) for various keys (b). KF = 4 for the key 'cntrl', KF = 2 for 'alt' and 'F'; KF = 1 for 'B', 'X', 'Z', and 'R'.

sequence 'Ctrl X' exists and the key-stroke 'Ctrl' is also used in three other commands, while the key 'X' occurs only in this single command, then the key 'Ctrl' has a fan degree of four (KF = 4) and the key 'X' has a fan degree of 1 (KF = 1).

The experiment examines the influence of task fan and key fan. An increase in the required processing time and in the number of errors committed is expected to occur as either task or key fan are increased.

Consistency

When investigating command languages, an important factor to be taken into consideration is consistency. Consistency as it pertains to human–computer interaction is analogous to meaningful connections among sentences in declarative knowledge. Consistency refers to the meaningfulness of command syntax and key assignments. High consistency reduces the effort of learning the word processing system. According to Payne (1989) consistency means that users 'categorise the world of tasks according to the similarities and differences among tasks, and they perceive higher-order rules which utilise the semantics of the task-world to capture structural resemblances between separate task-action mappings' (p. 138).

Another definition of consistency compatible to the above definition is possible. A system shows consistency when a cue is connected to the same operation for different tasks. Here a cue refers to any information element that is used to recall a user-operation. Consistency should result in positive transfer between tasks having the same cue-to-operation connection. This connection may be referred to as a *common element* among several tasks. Consistency will be illustrated with an example. An editor may possess two deletion commands, 'delete word' and 'delete character'. The first command requires the key-stroke-sequence 'Alt-W' and the second 'Alt-C'. The feature 'delete' is linked to the key 'Alt' for both commands. This connection represents a common element of both sequences.

Usually, consistency is confounded with the degree of fan as follows. If many different commands have a common element, e.g. a connection among a task feature and its assigned key, there will be positive transfer. However, this also increases the degree of fan, because an element (e.g. task feature) is now part of several commands, and is thus connected to multiple user actions. Consequently, consistency is expected to reduce the interference effects caused by fan. If the positive effects dominate interference may be entirely negated. Thus a sensible mapping of task and key components should serve to counteract the expected adverse effects of fan and perhaps even lead to increased learnability and performance.

Consistency within a command structure is analogous to the thematic organisation of sentences in sentence retrieval tasks, where fan-effects have been shown to be reduced by thematic coherence. Similarly, the interference of fan-effects in editing should be reduced by an increase in consistency. Therefore, the present investigation is also designed to examine the influence of consistency on interference.

Necessary experimental controls

An experimental study that is supposed to prove interference effects in a procedural domain requires that all sources which may overshadow interference be tightly controlled. Considerable effort is necessary to eliminate possible sources of confounding for the experimental conditions with various degrees of fan. As an important requirement, the degree of practice for commands with differing degrees of fan should be carefully matched. The evaluation of the data shouldn't rely on global measures, e.g. processing times for whole texts, but rather must apply at the level of single key-strokes (e.g. Singley and Anderson's 'Microanalysis', 1987–88). In order to take sources of variability (e.g. cursor positioning times) into account, the data should be analysed at the level of individual key-strokes. Attention to all these factors should aid in the identification of previously largely unnoticed interference effects in text editing.

Method

Subjects

Twenty-four students were recruited from the Technische Hochschule Darmstadt. All subjects, 18 female and 6 male students with an average age of 25, were editing novices without computer experience. They were randomly assigned to one of two experimental groups. All subjects participated on at least two successive days. On the first day, an introduction and learning stage of about half an hour was given, followed by the first half-hour editing session. On the second day, there were two half hour sessions. Twelve subjects, 6 from each experimental group, participated in two additional half-hour sessions on days 3 and 4. Subjects were paid DM 20.– for two, or DM 30.– for four days.

Overview and design

Subjects learned to use an editor that contained a set of cursor movement commands and 7 text modification commands. The subjects' key-strokes and key-stroke-latencies were recorded during several editing sessions.

There are four independent variables: editor consistency, degree of task fan, degree of key fan, and number of session. Half the subjects work with consistent editors and the other half with inconsistent ones. The task fan of the commands is varied within each editor through the levels of 1, 2, 4. The degree of task fan depends on the number of commands that have the same task component, either the same function or the same object, within an editor. The text modification commands require the subject to press two command keys in sequence. The degree of key fan depends on the number of commands that have the same first key-stroke. The second key-stroke was in each case unique. Key fan can take on values of 1, 2, or 4. For instance, a command having a key fan of 4 implies that there are three additional commands with the same first key-stroke. All subjects participated for at least three sessions. In each session, 15 texts had to be edited. Half of the subjects participated in two additional sessions.

Editor material

Twelve different editors were constructed with two subjects being assigned to each. Table 1 shows the selection of commands and keys for the editors. All editors included 6 cursor movement commands. They were 'move cursor one word to the right', 'move cursor one word to the left', 'move cursor one line up', 'move cursor one line down', 'move cursor to the first word of the line' and 'move cursor to the last word of the line'. These commands require two key-strokes. The first key indicates the type of cursor movement and the second one the direction.

Table 1

Selection of commands and keys for the experimental editors.

Editor no.	Inconsistent editors											
	Transpositions				Deletions				Insertions			
	T1 [a]	T2	T3	T4	D1	D2	D3	D4	I1	I2	I3	I4
1	OB_{44} [b]	OJ_{44}	IN_{42}	UL_{41}	–	IK_{12}	–	–	OH_{24}	–	–	OM_{24}
2	UN_{44}	IK_{42}	UJ_{44}	UB_{44}	UM_{24}	–	OL_{21}	–	–	IH_{12}	–	–
3	–	IM_{22}	OH_{24}	–	–	–	ON_{14}	–	OK_{44}	OB_{44}	UJ_{41}	IL_{42}
4	–	UN_{14}	–	–	–	OL_{21}	–	UJ_{24}	IK_{42}	IH_{42}	UB_{44}	UM_{44}
5	–	–	–	IB_{14}	IK_{44}	IM_{44}	UL_{42}	OJ_{41}	–	UH_{22}	IN_{24}	–
6	OL_{22}	–	–	UM_{24}	OB_{42}	UN_{44}	UH_{44}	UJ_{44}	–	–	–	IK_{11}

Editor no.	Consistent editors											
	Transpositions				Deletions				Insertions			
	T1 [a]	T2	T3	T4	D1	D2	D3	D4	I1	I2	I3	I4
7	UB_{44}	UJ_{44}	UN_{44}	UL_{44}	IM_{22}	IK_{11}	–	–	OH_{22}	–	–	OM_{22}
8	ON_{44}	OK_{44}	OJ_{44}	OB_{44}	–	–	IL_{22}	–	–	UH_{11}	–	–
9	–	UM_{22}	UH_{22}	–	–	–	IN_{11}	–	OK_{44}	OB_{44}	OJ_{44}	OL_{44}
10	–	UN_{11}	–	–	–	OL_{22}	–	OJ_{22}	IK_{44}	IH_{44}	IB_{44}	IM_{44}
11	–	–	–	IB_{11}	UK_{44}	UM_{44}	UL_{44}	UJ_{44}	–	OH_{22}	ON_{22}	–
12	OL_{22}	–	–	OM_{22}	IB_{44}	IN_{44}	IH_{44}	IL_{44}	–	–	–	UK_{11}

[a] Abbreviations: T1: header-word move-into-text, T2: header-word move-over-text, T3: header-word copy-into-text, T4: header-word copy-over-text, D1: delete character, D2: delete word, D3: delete rest of line, D4: delete entire line, I1: insert one character, I2: insert several characters, I3: overwrite one character, I4: overwrite several characters.

[b] Pairs of capital letters denote the two key strokes used for each of the 7 editing commands. The indices of the capital letters show the values of the independent variables: task fan and key fan. The first digit indicates task fan and the second, key fan. (e.g.: Editor 1 contains only one deletion command 'delete word'. This command requires the key stroke sequence 'IK' and has a task fan of 1 and a key fan of 2.)

Task fan

Each editor includes 7 text editing commands with the individual commands being divided into one of three groups, each comprised of commands similar in function (i.e. transposition commands, deletion commands or insertion commands).

In each editor, one group contains four commands, one group two commands, and the last group only a single command such that the desired total of 7 commands per editor is attained. The task fan is then defined as the number of similar commands within a specific command group in a specific editor. Thus, the task fan can take on values of 1, 2, or 4. The generation of each editor began with each of the command groups having four functions. The three command groups were then each assigned a unique task fan value of 1, 2, or 4 such that a total of 7 commands per editor was maintained. For the command group with a task fan of 4, all four of the original commands within that group were retained. The group assigned a task fan of 2 had two of its originally included commands randomly deleted. Similarly, the task fan 1 group was generated by randomly eliminating three of the original commands. This editor generation procedure ensures that the task fan is not confounded with the command type.

In commercially available editors, different commands require different execution schemes. Some have no cursor positioning requirements while others demand cursor positioning or even block marking. A prerequisite for the 12 commands selected for the experiment was, that the execution schemes within a single command group should be only as similar to one another as the schemes from commands of different groups are to one another. Otherwise positive transfer could arise within one command group due to similar editing schemes. This positive transfer could be larger for groups containing more commands. To avoid confounding the degree of fan with the similarity of the command-execution-scheme, the same simple editing scheme was chosen for all commands: First, the cursor had to be positioned in a way which required only one kind of cursor movement, and second, exactly two command key-strokes were required.

To satisfy the above demands, three fairly artificial command groups were constructed. These three groups were characterised as commands for 'deletion', 'transposition' and 'insertion'. The four deletion commands are 'delete character', 'delete word', 'delete rest of line' and 'delete entire line'. A transposition command takes a header word highlighted in a special status line and either copies or moves it into the text at the cursor. Copying implies that the header word is simply copied into the text, whereas moving implies that the header-word in the status line is cleared. In addition, the highlighted word can be either be inserted or it may overwrite the text at the cursor-position. Thus there are four different transposition commands defined; 'header-word move-into-text', 'header-word move-over-text', 'header-word copy-into-text' or 'header-word copy-over-text'. The insertion commands are 'insert one character', 'overwrite one character', 'insert several characters', 'overwrite several characters'.

Key fan

Different key fan degrees are provided. For four commands the same first key is used, two further commands are assigned another common first key and one command yet a different first key. All keys are selected such as to confine all key-strokes to a small area on the keyboard. For first keys 'U', 'I', and 'O' are used. The second keys are unique for each command. The keys 'B', 'J', 'H', 'K', 'L', 'M' and 'N' are employed.

Consistent vs. inconsistent editors

Within a consistent editor, commands within the same command set are assigned the same first key. Therefore the degree of key and task fan are always equal. Within inconsistent editors, the choice of the first key is not systematically related to any task feature; therefore key and task fan can be combined independently, as shown in table 1.

Text material

To keep the experimental conditions comparable, all commands had to be executed the same number of times. The complexity of the cursor movement is equal for all tasks. Because there are six types of editors that differ in their commands separate texts were developed for each of them. For one session, 15 texts are generated for each type of editor. The texts consist of word lists (see fig. 2). Each text includes the 7 tasks that belong to an editor. The sequence of tasks is chosen randomly for each text. In order for the subject to move from one editing task to the next, a single cursor command has to be applied at least once. Each of the 6 cursor commands occurs in random sequence once in each text. Tasks are not performed in the order of their appearance in the text. Coloured numbers from 1 to 7 indicate the position of the next task. For editing, the subjects are provided with a handout of the text with coloured correction symbols. The first task is always presented in the middle of the text (see fig. 2).

Editor learning and experimental procedure

Before taking part in the experiment, subjects received a short general introduction to editing. This was followed by a learning stage. Here subjects were presented a description of the commands and correction symbols. Each command was practised several times on the computer. In the case of questions or problems, an experimenter gave advice. Next, subjects had to practise all commands on an example text in random order. They received cards with pictograms and explanations of the correction symbols. The learning stage was completed when subjects were able to perform all 7 commands without errors twice consecutively.

In the actual experimental session, the text was presented on the monitor. In order to find where corrections were required, the subjects had to refer to a printout of the text pointing to the editing tasks and the execution sequence. Subjects were instructed to select the kind of processing that requires the smallest number of key-strokes. For instance, subjects were told to jump directly to the beginning of a line instead of moving the cursor wordwise. Also repeated applications of 'delete word' could not be used to delete a line. Each error or deviation from the optimal key sequence was followed by both an acoustic signal and on-screen error feedback. In this case, the command was not executed by the program.

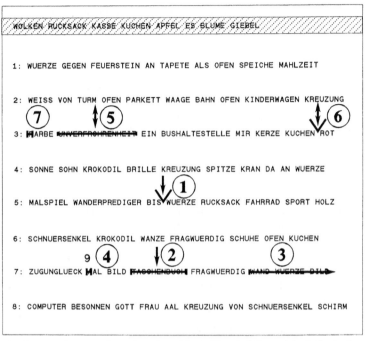

Fig. 2. Example of a text as presented to the subjects. The circled numbers indicate the required sequence of processing. In the original, correction symbols are coloured in red and task numbers are in blue. The header line contains the target words for the transposition commands. Task 1 was a 'header-word move-into-text', task 2 a 'header-word move-over-text', task 3 a 'delete rest of line', and in task 4 the digit 9 is substituted for the character below it, etc.

Hardware and programs

The experiment was carried out with two linked Apple IIe computers. The first of them evaluated the key-strokes, recorded their exact time and sent them, in case of agreement with the optimal sequence of key-strokes, to the second computer on which the editor ran. The required programs were written in Forth and in assembler code. The editors were written in C, and ran on an Apple IIe equipped with a 68000 processor-card. The editors operated at such speed that the subjects were unable to perceive any delay between their inputs and the program's response.

Computing dependent variables

The data consist of all key-strokes and latencies for each task. Each key-stroke is assigned a latency value computed as the difference between that key-stroke and the previous key-stroke. Deviations from the optimal key-stroke sequence are recognised as errors. The first of the 7 tasks included on a text is not evaluated. Also, the first text of each session is not used in the evaluation. Fourteen texts, each with 6 tasks, remain for evaluation. For each subject the arithmetic means of the latencies are computed for all experimental conditions. Only tasks with both command key-strokes correct are included. Command latencies of correct key-strokes are excluded, if either the first command key takes longer than 10 seconds or the second key longer than 5 seconds. This occurs in less than 1 percent of all tasks.

Results

The primary goal of this experiment is to show interference in using command languages. Interference is indicated by longer processing times and higher error rates. The task processing times are divided into two main components: First, cursor positioning, and second, proper execution of command keys. It is possible that command keys are recalled during cursor positioning. In this case the fan-effect may affect the cursor positioning component. If the command keys are recalled following the completion of cursor positioning, then only the command key execution component should be affected. For computing command-key latencies, the latencies of the two command keys are added together without cursor positioning times.

Unless otherwise indicated, all analyses of variance include two repeated-measurement factors. The first factor is a three-level session factor. Sessions 4 and 5 are excluded, because only half of the subjects went through them. The second factor is the three-level fan factor. For this experiment, key and task fan in inconsistent editors are not analysed together, because a number of the key fan/task fan combinations occur with too few subjects (see table 1). For example, only two subjects have both task fan and key fan equal to 1 (FAN_{11}). For this reason, key and task fan will be analysed separately in most of the following analyses of inconsistent editors.

Interference due to key fan in inconsistent editors

In inconsistent editors, the task components and keys are logically independent. Here, a key fan-effect is found in the expected direction for the command-key latencies (see fig. 3). This effect is statistically significant ($F_{2,22} = 4.86$, $p < 0.05$).

For cursor positioning, a key first had to be pressed that selected cursor movement, and then another key that indicated the direction of the movement. This direction key was pressed one or more times depending on the desired final cursor position. To keep the number of cursor-positioning key-strokes constant for analysis purposes, only the latencies of the cursor-selection key and the first press of the direction key are summed. With growing degrees of fan, the cursor-positioning latencies show only a relatively small increase (approx. 200 ms) as compared to the command-key latencies (approx. 800 ms). The increase is confined to the first segment of the fan function (see fig. 4), and the overall effect is not statistically significant ($F_{2,22} = 1.21$, $p > 0.05$).

On average, less than 3% of the responses were errors and none of the effects on accuracy proved statistically significant. There is, however, a systematic trend in the data for the errors to increase with the degree of fan (see table 2). Many experiments

Execution time [ms]

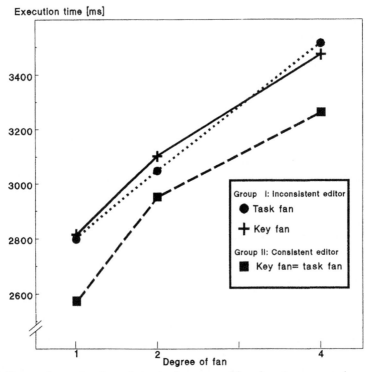

Fig. 3. Command execution times. Latencies for command keys have been averaged over the first three sessions. For consistent editors, the two types of fan are indistinguishable.

using sentence recognition report significant increases in reaction time with increasing degrees of fan. Most of these also fail to validate the effects on response accuracy (e.g. Anderson 1974).

To summarise, in inconsistent editors the use of the same key in different commands leads to interference. The more commands that share a common key-stroke, the more time needed to execute those commands. This effect is statistically significant for command-key latencies but not for cursor-positioning latencies.

Interference due to task fan in inconsistent editors

A task fan-effect is found for command latencies (see fig. 3). It proves to be significant in a two-factor analysis of variance with the factors SESSION and TASK FAN ($F_{2,22} = 4.21$, $p < 0.05$). Similar to the results concerning key fan, the task fan-effect can't be obtained for errors (see table 2) and cursor positioning latencies (see fig. 4; $F_{2,22} = 1.99$, $p > 0.05$). It should be noted, however, that there is again a trend towards fan-effect. In summary, the observed task fan-effects show, that in inconsistent editors, similar task-meaning components lead to interference, as was predicted.

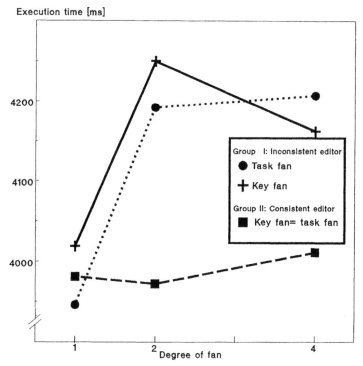

Fig. 4. Cursor positioning times. Latencies for cursor keys have been averaged over the first three sessions.

Table 2
Percentage of execution errors for command keys averaged over the first three sessions.

| | Degree of fan | | | | | | | |
| | First command key-stroke | | | | Second command key-stroke | | | |
	1	2	4	Mean [a]	1	2	4	Mean [a]
Inconsistent editors								
Task fan	1.0	2.1	4.9	3.5	2.8	1.1	2.5	2.1
Key fan	4.0	3.7	3.3	3.5	0.7	0.9	3.1	2.1
Mean	2.5	2.9	4.1	3.5	1.8	1.0	2.8	2.1
Consistent editors								
Task fan = Key fan	2.4	2.1	2.2	2.2	0.0	1.0	1.6	0.9

[a] Simple means are weighted by command frequency.

Fan-effects in consistent editors

In consistent editors where meaningful assignments exist between task features and the keys required for command execution, it is not possible to distinguish between task fan and key fan. Here, necessarily, the task fan degree equals the key fan degree. The fan-effect for command latencies was also found with consistent editors. An increase in fan leads to longer command latencies (see fig. 3). An analysis of variance with repeated measurement on the factors SESSION and degree of FAN shows a significant effect for degree of FAN ($F_{2,22} = 5.09$, $p < 0.05$).

Comparison of fan-effects among consistent and inconsistent editors

It was assumed that latencies in consistent editors would prove to be shorter than in inconsistent ones. Fig. 3 also shows that the subjects tend to work faster with consistent editors than with inconsistent ones. Unfortunately, this main effect of consistency cannot be proved to be statistically significant in two analyses of variance with one between-groups factor (type of EDITOR) and two repeated-measurement factors (FAN and SESSION). The first analysis compared the consistent editors to the inconsistent ones with the factor TASK FAN (Factor EDITOR: $F_{1,22} = 1.61$, $p > 0.05$). The second analysis involved the factor KEY FAN (Factor EDITOR: $F_{1,22} = 1.08$, $p > 0.05$).

If consistency is based on common cue-operation.connections for different task as defined previously, consistency will not be separable from fan. For the consistent editors, positive transfer between different tasks should rise with increasing degree of fan. Therefore fan-effects should be reduced in consistent editors as compared to inconsistent ones. The results as presented in fig. 3 are, however, not suitable for the comparison between interference effects arising in consistent and inconsistent editors. The figure does not reflect the fact that for the inconsistent editors, either the task or the key fan rises from a degree of 1 to 4, while in consistent editors both factors rise simultaneously. The three combinations FAN_{11}, (first index: degree of task fan, second index: degree of key fan), FAN_{22}, and FAN_{44} should be compared with the three corresponding combinations of the inconsistent editors. A direct comparison, however, is not possible, because there are too few subjects for these fan combinations in consistent editors (see table 1 and table 3).

Consequently, this data was combined to form the two levels 'Both-fan-low' (BFL) and 'Both-fan-high' (BFH). The values of the combination FAN_{44} are taken as 'Both-fan-high' values and FAN_{11} values are taken as 'Both-fan-low'. If FAN_{11} is missing for a subject, then the mean of FAN_{12} (task fan = 1 and key fan = 2) and FAN_{21} (task fan = 2 and key fan = 1) is taken. So, for 8 subjects on the inconsistent editors Both-fan-low values are provided. The increase in latencies from Both-fan-low to Both-fan-high is nearly twice as large for inconsistent editors as for consistent ones (BFL = 2523 ms and BFH = 3722 ms vs. BFL = 2569 ms and BFH = 3262 ms). Note that this is possibly a conservative estimate because inconsistent editors have higher fan degrees for Both-fan-low values than consistent editors do. An analysis of variance with factors FAN (BFL vs. BFH), SESSION (1, 2, 3) and type of EDITOR (consistent vs. inconsistent) shows a statistically significant difference in the increase of inter-

Table 3
Command key-stroke latencies (in ms) for consistent and inconsistent editors for the first three sessions.

Degree of task fan	Degree of key fan					
	Inconsistent editors			Consistent editors		
	1	2	4	1	2	4
1	*2265 (2)* [a]	2666 (4)	3067 (6)	*2569 (12)*	–	–
2	2309 (4)	*3151 (6)*	3159 (12)	–	*2951 (12)*	–
4	3329 (6)	3313 (12)	*3722 (12)*	–	–	*3262 (12)*

[a] The values in parentheses give the number of subjects assigned this fan combination. Means that are comparable between both groups are italicized.

ference between the two types of editors (Interaction FAN × EDITOR: $F_{1,18} = 7.17$, $p < 0.05$). This can be interpreted as showing that interference due to fan is reduced in consistent editors as compared to inconsistent ones.

Separate analysis of the first and second command key-strokes

The latencies of each of the two command key-strokes can be analysed separately. This serves to investigate the temporal relationship between the retrieval of command information from memory and the execution of key-strokes. Task fan and key fan should have different effects on the recall of the first and second command keys. Task features provide information for the recall of both command keys, therefore the task fan should have an effect on both key-strokes. The first command key restricts the number of second key-stroke alternatives. Thus, the key fan of the first command key should influence the recall of the second command key but should not influence the recall of the first command key itself.

Two types of relations between memory recall and user operation may be possible, the first command key could be pressed following the complete recall of both keys from memory, or, the motor program for the first key-stroke could be started before the second key is recalled. If the latter were the case, an interaction between TYPE-OF-FAN and KEY-STROKE-ORDER should arise. A tendency towards this interaction is

Table 4
Increase of latency (in ms) from fan degree 1 to fan degree 4.

	First command key-stroke	Second command key-stroke
Inconsistent editors		
Task fan	583	132
Key fan	349	313
Consistent editors		
Task fan = Key fan	451	242

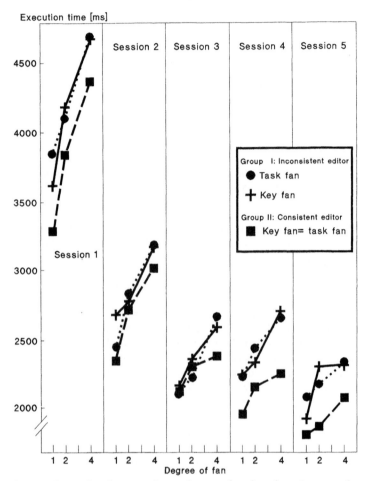

Fig. 5. Command execution times over five sessions as a function of consistency and type of fan.

found in the data (see table 4), but it is not statistically significant ($F_{1,11} = 1.18$, $p > 0.05$) in an analyses of variance for inconsistent editors with the factors TYPE-OF-FAN (task fan vs. key fan), KEY-STROKE-ORDER (first vs. second command key-stroke) and SESSION (1,2,3). The factor KEY-STROKE-ORDER ($F_{1,11} = 5.01$, $p < 0.05$) is statistically significant, indicating higher increases in latency for the first than for the second command key-strokes.

For consistent editors it is expected that the interference should be low for the first and high for the second command key-stroke, because the recall of the first command key requires an information-operation association that is always the same for a given type of task (e.g. delete). Therefore positive transfer should reduce interference especially for the first command key-stroke. Users may first identify the type of task (e.g. a deletion task) and then begin to press the appropriate first command key before recall of the second command key, where further specific information is needed. However, this effect is not obvious in the data, when consistent and inconsistent editors are compared (see table 4).

Summarising, the data speak for unitary recall and execution of commands, although in inconsistent editors a tendency towards different effects on the two command key-strokes is found, as structural differences between task and key fan would predict.

Practice induced changes in task and key fan-effects

Studies in which sentences are learned, generally show a reduction but not an elimination of fan-effects with practice. The data from the present experiment also show a decrease of fan-effects from one session to the next (see fig. 5). The decrease is found for both key and task fan-effects, however both fan-effects may still be noted even in the last session.

In order to calculate the change in interference, the difference between Both-fan-low and Both-fan-high values were used. These provide extreme combinations of fan degrees that are especially powerful for demonstrating the existence of fan-effects. To show the reduction of latency differences between Both-fan-low and Both-fan-high, the contrast between sessions 1 and 3 was calculated. [1] In the analysis of variance with the factors EDITOR (consistent vs. inconsistent), SESSION (1 vs. 3), FAN (BFL vs. BFH) the expected interaction between SESSION and FAN ($F_{1,18} = 15.84$, $p < 0.01$) was found. Despite the reduced number of subjects participating in session 5, the Both-fan-low and Both-fan-high values from that session still show significant fan-effects. An analysis of variance with the factors EDITOR and FAN shows them to be statisically significant during the fifth session ($F_{1,8} = 8.82$, $p < 0.05$).

Discussion

In the experiment two different types of interference effects are demonstrated. Interference occurs between the commands associated with similar tasks. This is called task fan. Furthermore, interference is also evident when different commands require the same key-strokes. The latter is referred to as key fan-effect. Both results can be explained by the fanning of 'information elements' to several different user actions. For task fan-effects, different features of the editing task show

---

[1] Session 3 was chosen because all subjects were present, whereas in sessions 4 and 5, only half of the subjects took part.

different degrees of fan. For key fan-effects, keys are applied by the user that have different fan degrees and furnish information for the retrieval of the next user action.

The controversy reflecting declarative and procedural knowledge

In the ACT* theory of Anderson (1983) fan-effects only affect declarative knowledge. According to Anderson (1987) editing requires primarily procedural knowledge for which fan-effects should play little or no role. The 'identical production theory' of Singley and Anderson claims a 'lack of negative transfer' (1989: 223) for skills such as text-editing, calculus, and programming in a LISP tutor. 'These predictions were supported in a study examining transfer between two text editors designed for maximal interference' (1989: 223). This citation refers to their experiment (1987–88) where editing novices learned an EMACS compatible editor for two sessions, then switched to an editor with completely reassigned key-bindings for two sessions, and then returned to the first editor. By way of comparison, sessions took longer in Singley and Anderson's experiment than in ours, but more commands had to be learned in their study. Differences in the results are unlikely to be explained with different degrees of learning. The reason for finding systematic interference in this experiment, but not in Singley and Anderson's study, probably lies in the different experimental designs. In Singley and Anderson's study, the same commands were practised with new key assignments after switching to the second editor. Thus, interference due to new key assignments was likely obscured by positive transfer due to common elements. Common elements result from the fact that all commands in both editors were the same except for their names.

One goal of this investigation was the explicit comparison between interference in editing and interference in a declarative domain such as sentence learning. This experiment shows interference effects that are quite similar to fan-effects of experiments with sentence retrieval, and that seem to depend on the same variables. The increase of latencies due to fan was shown to be significant. Also, substantial properties of the propositional fan-effect emerge in the editing data:

(1) The increase of latencies is largest for a change of fan degree from 1 to 2. The greater increase of fan from 2 to 4 yields a quantita-

tively comparable and thus relatively small increase in mean latencies. Similarly, in experiments with sentence retrieval, the main increase of latencies is also observed between the degree of 1 and 2 (e.g. Aschermann and Andres 1990; Heydemann 1989: 64).

(2) The amount of the fan-effect is reduced by practice, but is still apparent even after five editing sessions. The same result has been found in sentence retrieval experiments (e.g. Pirolli and Anderson 1985).

(3) Consistency moderates the fan-effects in editing. Similar results have been reported for sentence retrieval, where thematically coherent episodes were learned (Smith et al. 1978).

(4) The fan-effects in editing were significant with latencies but not with error rates. Also in sentence retrieval experiments, fan-effects are easier to prove significant by using latencies than by using error rates (e.g. Heydemann 1989: 38).

The results of this experiment clearly argue against the classification of editing as strictly procedural knowledge with no declarative knowledge components. Either declarative and procedural knowledge are not so distinct as suggested by Anderson (1987), or, both procedural and declarative knowledge are relevant to editing.

Differences and similarities between editing and sentence retrieval

The knowledge representation of editing commands may be assumed to be comparable to the representation of items which are used in paired-associates tasks or in sentence learning. The key-strokes of an editor command correspond to the answer in a paired associates or sentence retrieval task. The editing task and the keys that the user has already recalled represent the stimuli for the remaining responses. On the other hand, there are also several differences between editing and paired associates or sentence retrieval tasks. One difference is that the stimulus onset is not observable for the task of editing. This seems however to be a mere methodical difference. There is a more crucial difference concerning the cursor positioning. Cursor positioning fulfils the role of a secondary task during which the following command key-strokes may be retrieved from memory, thus potentially obscuring interference effects. Nevertheless, the data in this experiment succeeded in showing fan-effects. The most important difference between text editing and paired associates or sentence retrieval tasks may result from the increased complexity of editing. The execution of an editing task requires many steps, each of which may either lead to interference or, conversely, to positive transfer. If interference occurs at one step, another step may show positive transfer overshadowing interference. This is probably the main factor responsible for reducing the amount of observable interference in editing experiments (e.g. Singley and Anderson 1987–88). In our experiment a great deal of care went into the prevention of this confounding. Also the data were analysed at the level of single key-strokes. Interference effects would have been masked, if the analysis had been carried out on the task as a whole.

Consistency

One of the main topics of the experiment was the investigation of the interaction among editor consistency and interference due to fan. Consistency results from a correlation between task features and user actions. There is a logical relationship between degree of fan and consistency. If the same task features in different commands are connected to the same user actions, the degree of consistency is high. At the same time, if there are identical task features in different commands, then their degree of fan is correspondingly increased. Thus, consistency is normally correlated with the degree of fan. Therefore increased consistency should reduce the interference associated with high degrees of fan. This interpretation is supported by the data of the experiment.

Consistency arises from associations of an information cue to the *same* user operation in different tasks. The close connection between interference and positive transfer is emphasised in this definition, because a same information cue connected to *different* user operations leads to interference. Theories for command language systems (e.g. the Cognitive Complexity Theory of Kieras and Polson (1985) or the Task Action Grammar of Payne and Green (1989)) that only consider the sources of positive transfer should be extended to take interference into account.

Implications for practical use

In addition to the theoretical consequences that result from the existence of fan-effects in editing, there are also consequences of practical relevance. The experiment reveals the danger of providing a system with too many similar functions. Particularly, when such systems are not sufficiently consistent, interference may counteract the benefits of greater functionality.

A further consequence of the experimental results relates to the action sequences required of the user. It may not be useful to provide systems with a rich variety of commands with fanned components unless consistency is high. For instance, if consistency rules are not obeyed, the use of the same function key in connection with several control keys (e.g. 'Alt', 'Shift', 'Ctrl', 'Escape') should lead to high interference. An example of reasonable keybindings is suggested by Walker and Olson (1988), who recommend that the 'Esc' key be used for all system-related commands, the 'Alt' key for all deletions and the 'Ctrl' key for all other commands.

Often, it may not be possible to build a highly consistent system or a system where each command has its own key. If many different functions without sufficient consistency rules have to be mapped to key sequences, the choices of the degrees of key fan are important. For fan-effects the main increase of interference occurs between fan degrees of 1 and 2. Therefore, it would be better to choose high fanning for a few keys and none for the others. Medium fan degrees for all keys might prove inferior. For commands that are of major importance, either relatively short key sequences or single key-strokes should be used.

However, in practice, decisions may often be difficult because a design decision for a system may simultaneously affect factors influencing interference and positive transfer. In these cases, a carefully performed analysis that takes all factors into account, may not provide conclusive suggestions but only hints. As a rule, it always appears to pay, if the consistency of the system is increased. Then the fan-induced interference might be reduced even if the user doesn't explicitly recognise the applied consistency rules. However, the amount of positive transfer may be considerably increased if the common elements that produce consistency are not only implicitly but explicitly perceived by the user. This is suggested by experiments with sentence learning of Whitlow et al. (1982). They showed that subjects who received an appropriate instruction for explicitly paying attention to common elements in the text material had reduced fan-effects compared to an uninstructed control group. This suggests that system design should not only implement consistency, but also explicitly point out to the user the regularities built into the command system.

References

Anderson, J.R., 1974. Retrieval of propositional information from long-term memory. Cognitive Psychology 6, 451–474.

Anderson, J.R., 1976. Language, memory, and thought. Hillsdale, NJ: Erlbaum.

Anderson, J.R., 1983. The architecture of cognition. Cambridge, MA: Harvard University Press.

Anderson, J.R., 1987. Skill acquisition: Compilation of weak-method problem solutions. Psychological Review 94, 192–210.

Aschermann, E. and J. Andres, 1990. Aspekte des Fächer-Effektes: Linearität und Vorwissenseinfluss. Zeitschrift für experimentelle und angewandte Psychologie 37, 1–15.

Card, S., T.P. Moran and A. Newell, 1983. The psychology of human–computer interaction. Hillsdale, NJ: Erlbaum.

Crowder, R.G., 1976. Principles of learning and memory. Hillsdale, NJ: Erlbaum.

Green, T.R.G. and S.J. Payne, 1984. Organisation and learnability in computer languages. International Journal of Man–Machine Studies 21, 7–18.

Grudin, J. and P. Barnard, 1984. The cognitive demands of learning and representing command names for text editing. Human Factors 26, 407–422.

Heydemann, M., 1989. Der Fächerungreffekt: Experimente zur Gedächtnisinterferenz und ein konnektionistisches Modell. Unpublished Dissertation, Darmstadt, Technische Hochschule.

Jones, W.P. and T.K. Landauer, 1985. Context and self-selection effects in name learning. Behavior & Information Technology 4, 3–17.

Karat, J., L. Boyes, S. Weisgerber and C. Schafer, 1986. 'Transfer between word processing systems'. In: M. Mantel and P. Orbeton (eds.), Proceedings of the CHI '86 Conference on Human Factors in Computing Systems. New York: ACM. pp. 67–71.

Kieras, D.E., 1988. 'Towards a practical GOMS model methodology for user interface design'. In: M. Helander (ed.), Handbook of human–computer Interaction. Amsterdam: North-Holland. pp. 135–157.

Kieras, D. and P.G. Polson, 1985. An approach to the formal analysis of user complexity. International Journal of Man–Machine Studies 22, 365–394.

Landauer, T.K., K.M. Galotti and S. Hartwell, 1983. Natural command names and initial learning: A study of text-editing terms. Communications of the ACM 26, 495–503.

Myers, J.L., E.J. O'Brien, D.A. Balota and M.L. Toyofuku, 1984. Memory search without interference: The role of integration. Cognitive Psychology 16, 217–242.

Osgood, C.E., 1949. The similarity paradox in human learning: A resolution. Psychological Review 56, 132–143.

Payne, S.J., 1989. 'A notation for reasoning about learning'. In: J. Long and A. Whitefield (eds.), Cognitive ergonomics and human–computer interaction. Cambridge: University Press. pp. 134–165.

Payne, S.J. and T.R.G. Green, 1989. The structure of command languages: An experiment on task-action grammar. International Journal of Man–Machine Studies 30, 213–234.

Pirolli, P.L. and J.R. Anderson, 1985. The role of practice in fact retrieval. Journal of Experimental Psychology: Learning, Memory, and Cognition 11, 136–153.

Polson, P.G., S. Bovair and D.E. Kieras, 1987. 'Transfer between text editors'. In: J.M. Carroll and P.P. Tanner (eds.), CHI + GI Conference Proceedings: Human Factors in Computing Systems and Graphics Interface. New York: ACM pp. 27–32.

Polson, P.G. and D.E. Kieras, 1985. 'A quantitative model of the learning and performance of text editing knowledge'. In: L. Bornham and B. Burtis (eds.), Proceedings of the CHI '85 Conference on Human Factors in Computing Systems. New York: ACM. pp. 207–212.

Polson, P.G., E. Muncher and G. Engelbeck, 1986. 'A test of a common elements theory of transfer'. In: M. Mantel and P. Orbeton (eds.), Proceedings of the CHI '86 Conference on Human Factors in Computing Systems. New York: ACM. pp. 78–83.

Reder, L.M. and J.R. Anderson, 1980. A partial resolution of the paradox of interference: The role of integrating knowledge. Cognitive Psychology 12, 447–472.

Reder, L.M. and B.H. Ross, 1983. Integrated knowledge in different tasks: The role of retrieval strategy on fan effects. Journal of Experimental Psychology: Learning Memory, and Cognition 9, 55–72.

Schmidt, R., E. Fischer, M. Heydemann and R. Hoffmann, 1991. Searching for interference among consistent and inconsistent editors: The role of analytic and wholistic processing. Acta Psychologica 76, 51–72.

Shneiderman, B., 1987. Designing the user interface: Strategies for effective human–computer interaction. Reading, MA: Addison-Wesley.

Singley, M.K. and J.R. Anderson, 1985. The transfer of text-editing skill. Journal of Man–Machine Studies 22, 403–423.

Singley, M.K. and J.R. Anderson, 1987–1988. A keystroke analysis of learning and transfer in text editing. Human–Computer Interaction 3, 223–274.

Singley, M.K. and J.R. Anderson, 1989. The transfer of cognitive skill. London: Harvard University Press.

Smith, E.E., N. Adams and D. Schorr, 1978. Fact retrieval and the paradox of interference. Cognitive Psychology 10, 438–464.

Streitz, N.A., W.A.C. Spijkers and L.L. van Duren, 1987. 'From novice to expert user: A transfer of learning experiment on different interaction modes'. In: H.J. Bullinger and B. Shackel (eds.), Human–computer interaction – Interact '87. Amsterdam: Elsevier. pp. 841–846.

Waern, Y., 1985. Learning computerised tasks as related to prior task knowledge. International Journal of Man–Machine Studies 22, 441–455.

Walker, N. and J.R. Olson, 1988. Designing keybindings to be easy to learn and resistant to forgetting even when the set of commands is large. Proceedings of the Conference on Human Factors in Computing Systems, CHI '88. New York: ACM. pp. 201–206.

Whitlow, J.S., Jr., E.E. Smith and D.L. Medin, 1982. Retrieval of correlated predicates. Journal of Verbal Learning and Verbal Behavior 21, 383–402.

Ziegler, J.E., H.U. Hoppe and K.P. Fähnrich, 1986. 'Learning and transfer for text and graphics editing with a direct manipulation interface'. In: M. Mantel and P. Orbeton (eds.), Proceedings of the CHI '86 Conference on Human Factors in Computing Systems. New York: ACM. pp. 72–77.

Acta Psychologica 78 (1991) 287–304
North-Holland

On the microstructure of learning a wordprocessor *

Yvonne Wærn

Stockholm University, Stockholm, Sweden

This paper aims at studying the microstructure of computer users' actions when learning a wordprocessing system. Latencies for different actions, i.e. identifying task, planning method, selecting command, typing argument, closing command and evaluating result, during the performance of editing tasks were registered. It was found that the learning curves for latencies for all actions except identifying task, could be fitted by a power function, with a power constant around 0.90. The fit was good and the constants similar for two different commands (delete and insert). Also for the different actions observed the power constants were similar, although the curves started at very different points, ranging from 4.31 sec. to 21.72 sec. Further, a model was proposed for basic operations behind the observed actions. The model proposes a sequential procedure, where planning and recalling precede action. This model was tested against two different commands and was found valid for all data on the 2% level for the delete command and on the 8% level for the insert command. The model was also tested for individual subjects, where it was found that it was adequate for at least 80 percent of the subjects in three out of four cases tested. A closer study of the model fit suggested that the recall of a command was often performed in parallel with the planning of a method. Three subjects refuted the model by reflecting *after* their actions instead of *before* as proposed by the model.

Introduction

This paper aims at studying the microstructure of computer users' actions when learning a computer system.

* The research has been supported by a grant from the Swedish Council for Research in the Humanities and Social Sciences. I am indebted to Lars Rabenius for constructing the system which registered the data and which identified the subjects' keystrokes. Lars Rabenius and Robert Ramberg ran the experiment. I am also indebted to Nils Malmsten who performed the statistical calculations.

Author's address: Y. Wærn, Dept. of Psychology, Stockholm University, 106 91 Stockholm, Sweden.

Approaches to studying microstructure

Some different approaches towards studying the microstructure have been suggested. One derives from observing the time taken to perform different tasks during the use of a computer system. The elements consist of actions, which can be time-stamped during their execution. This approach is derived from the GOMS model (Card et al. 1983) and has most often been used to predict skilled behaviour in a system with specified requirements.

Another approach consists in deriving a set of rules, underlying the performance of the task. The rules are formulated as production rules. This is the so-called 'Cognitive Complexity Theory' or CCT (Polson and Kieras 1985), and has most often been used to predict learning of a certain system. The more rules to be learnt, the longer the time for learning.

Both these approaches merit further analyses. The GOMS model is designed to be an approximate model. Since only execution time can be observed, Card et al. estimated the mental times taken by obtaining independent measurements of the physical operations. The remaining time is the time taken for the 'mental operator'. This time comprises defining goals, planning and selection of methods. In one experiment, reported by Card et al., it was found that 60% of the task (manuscript editing) was mental time and 22% only was spent in typing. This estimation challenges us to get further into the analysis of the mental operator(s).

The CCT approach is approximate to the extent that the basic definition of what should count as a single production is arbitrary. This means that a calculation based on the number of productions suffers if the basic elements of the productions are not adequately specified. Since the production consists of an observation side and a performance side, a natural analysis would be to identify more basic elements on both sides. It has been shown that not only the number but also the content of productions affect the transfer between different situations. In particular, the direction of transfer depends on the relation between the old and the new situation on the observation side as well as on the performance side. When both are similar, positive transfer will be ensured (cf. Singly and Anderson 1987–88). When only the observation side is similar, whereas the performance side differs, negative transfer has been found (cf. Wærn 1985).

Here we want to attempt a search for some 'basic' elements, which can be used as a basis both for suggesting basic elements to enter into the productions and as a basis for analyzing the crude 'mental operator' as suggested by Card et al. (1983). Since the CCT approach is derived from the GOMS model it was considered most appropriate to start with a GOMS-type analysis. In order to get into the microstructure, the lowest level possible of the family of GOMS models was chosen, i.e. the keystroke level.

The keystroke-level model was first introduced by Card et al. (1983), who propose it to be 'useful where it is possible to specify the user's interaction sequence in detail'.

Card et al. suggest the execution time to consist of four physical-motor operators, K (keystroking), P (pointing), H (Homing) and D (drawing), one mental operator M, and a system response operator R, as in the following expression:

$$T_{execute} = T_K + T_P + T_H + T_D + T_M + T_R.$$

Given that a thorough task analysis is performed, where the method to perform the task can be specified in detail, and the execution times for each of the operators involved are independently measured, the time required to perform a given task can be predicted. Card et al. report that the model could predict performance over a variety of different tasks and systems with a standard error of 21%.

In order to get deeper into the investigation of the mental operator, a task analysis, suggesting a hypothetical procedure for performing the task will first be performed and further related to a possible empirical analysis.

Assumptions behind the task analysis

The tasks to be approached here are, as in Card et al., simple editing tasks where the system is command based. Since the task analysis is to a great extent concerned with mental tasks, certain assumptions have to be made.

Firstly, the GOMS model requires assuming that the activities can be described in a hierarchical way. The general goal for the task can be broken down into subgoals, which can be further broken down into methods to achieve these subgoals, and finally the methods can be broken down into physical actions. This means that there will be no recursions, such as changing goals or plans after the subject has selected the method or action. This means that events where errors and restarts have been made will not be considered in the data treatment.

Secondly, the GOMS model proposes the activities are made up of operations which are performed in sequence. The time to perform a whole procedure can be analyzed as a sum of the time taken for the more basic operations. The reason for this is that some activities simply have to follow each other. For instance, the user cannot plan which method to use in order to perform a task before she has identified the task. She cannot choose between methods before she has retrieved the methods which are feasible. She may retrieve the name of the command(s) to be used for a particular method in parallel with retrieving the method. The command cannot be executed until its name has been retrieved, however, and the result cannot be evaluated until the activity has been performed. Of course, if is possible to find iterative circles of task identification, method planning, command retrieval, action and evaluation of result. Such iterations take place when errors have been made or the subject by some other reason is not quite satisfied with the result achieved. For each error-free result derived, a sequential course of some operations can be found, however.

Thirdly, it will be assumed that although some mental activity can take place in parallel with some motor activity, the time taken for the mental operation will be constant under different motor activities. The assumption of a constant mental time (if any) in parallel with motoric activities must be made in order to make sense out of the latencies. This was not spelled out by Card et al., but their calculations show that they either considered the mental time occupied in parallel with the physical times to be negligible or constant.

The task analysis

The task used was, as in the case of Card et al., a simple editing task. The system is purely command-driven, so that only keystrokes (no mouse-moves) were concerned. The situation was designed so that the user got a text containing typing errors, on the screen. On each line of the text, there was one error. The correct spelling was placed to the left of the line. Thereby the goal was well specified. The subject could only perform one task (line) at a time, whereby the size of the unit task was predefined. The task analysis is aimed at describing the steps to be made in order to perform this task.

The system required the user to move the cursor to the place where the correction will be made. Then a command key had to be pushed in order to start the command mode. Then the proper command had to be given, with a possible argument. After this, the command had to be finished by pushing another key, and finally, a new task could be ordered by pushing still another key. This procedure is further analyzed in table 1, which also indicates how the part actions were time-stamped.

This task analysis is strictly related to the observations performed. Now the search starts for some more basic operations lying behind the observed latencies. This will be done in a hypothetical-deductive way. A model is proposed which suggests that the time taken for each observed activity is a linear function (preferably a simple sum) of the latencies for some basic operations. These basic operations may recur in the different observed activities. The model will assume a particular sequence of the basic operations, to fit into the sequence of the observed activities. The main assumption of the model is that people think first and act then. This means that the model suggests that operations requiring reflection (planning and recalling) are only involved in the first few observed actions. After the physical actions (type argument and close command) the model suggests that only a simple check for correctness follows, and no further reflections. The details of the model are given in table 2.

The reasons for these assumptions are the following. In the first activity observed (i.e. identify task), we suppose that the user reads the task and starts planning how to approach the task before starting to move the cursor. This planning can either concern how to get to the position where the correction shall take place or consist in high-level planning. The results of some researchers (Robertson and Black 1986) indicate that only short sequences of actions are planned at a time. Whatever is the case, the model assumes that the time for each planning step is constant.

When the user has arrived at the point where the correction shall be made, it is suggested that first the method(s) which are possible to use for this task will be retrieved from long-term memory. If several methods are considered, the retrieval may take proportionally longer. It is assumed, however (in order not to have more basic operations than actions), that only one method is retrieved. This is caught by the 'recall'

Table 1

A task analysis of editing in the situation and system used.

Proposed action	Corresponding observation
1. Identify task The first step consists of reading the correction to be made and finding the place in the text where the error is placed. During this step the plan for performing the task is identified at a high level. (See further below.)	1. Between acquiring task and starting the cursor movement.
2. Cursor movement The subject has to move to the place where the text should be corrected. The time taken is mainly motoric, and depends on the distance to travel.	2. The time taken from the start to the end of the cursor movement. This time was difficult to use because the cursor key was repetitive whereby some very short latencies were obtained.
3. Plan method When the cursor is at the right place, the plan started at the first step must be further developed by a concrete planning of the method to be performed. The planning of the method can have been started already during the cursor movement.	3. From the last cursor press to the pushing of the command starter.
4. Select command At this step a further specification of the plan is made by selecting the actual command to be used. The selection of a command is to some extent included in the method selection in step 3. The extra time taken here includes: a specification of the method, a mental repetition of the command to be given and the finding of the command on the keyboard.	4. From the control key to the typing of the command letter. (Only one-letter commands were used.)
5. Select argument The argument to the command is here selected and typed in. It can either be a figure or a text string.	5. From the command to the end of the argument. (If a text string was given, the measurement was based on an average over the letters typed.)
6. Close command After typing the command argument, the subject must close the command. Before closing, she has the chance to change the command and argument given.	6. Between the end of the argument and the closing of the command.
7. Evaluate result Finally, the result of the correction made is evaluated before the subject acquires next task.	7. From the closing of the command to the ordering of next task by pushing the down arrow key.
8. Error correction Errors can be corrected in the middle of writing commands or arguments or by an 'undo'-command.	8. Only error-free data will be used in the time analyses.

Table 2
Suggested mental operations involved behind the different actions observed.

Activity observed	Operations			
Identify task	Read	Plan		Type
Plan method		Plan	Recall	Type
Select command			Recall	Type
Select argument				Type
Close command				Type
Evaluate result	Read			Type

operation. Secondly, the application of one method to this particular situation has to be planned. The time taken for planning the method is supposed to be the same as for the first planning of the task, as mentioned above.

Thirdly, the name of the command has to be retrieved from long-term memory. As a first approximation it is suggested that this takes as long as retrieving the method to be used. This operation is thus also covered by the 'recall' operation.

The typing of the argument does not call for any retrieval from long-term memory, nor any planning. It is here suggested that everything is already planned at least at the command stage, so that only typing remains to be done.

The closing of the command is supposed to be an automated action, lying within the same time range as the typing. Thus the time taken for typing will be relevant for this activity.

Finally, before approaching the next task, the user has the chance to review what she has done, which involves reading in the same way as when identifying the task. No further thinking is supposed to be involved after the action has been performed. This is due to the main idea of the model mentioned above, that people think first and act later.

All activities involve the primitive operation of 'typing', which can be taken as a constant basic element. (This was also done in Robertson and Black 1986.)

Questions to be focused

Two general questions will be asked in this investigation:

(1) How can the learning of the system be described in terms of the observed part activities?
(2) How does the model for more basic operations fit the data?

Editing method to be focused

Since so many possibly confounding factors may affect time measurements, it was decided to focus *one method* only. In order to get stable enough data, this method would be the dominant one in the study. However, in order to be able to study the effect of having other methods available, other methods would also be taught and encouraged. The editing method focused was decided to be the insert/delete method, which allowed for some test of the generalizability of the model by including two different commands: insert and delete respectively.

Method

Situation

The subjects were given a text covering a couple of screen pages. At each line of the text there was a typing error to be corrected. The correct text was placed to the left of the line. The subjects could only correct one line at a time. The next task was not available until the first one was performed correctly. However, the whole screen page was visible to the subject.

The system

The wordprocessing system used was a simulation of a simple full-screen editor, using only commands. In order to invoke the command-mode, the subjects had to press the control key. In order to exit the command-mode the subjects had to press the return key. The cursor was moved with arrow keys. When the control key was pressed, a 'window' opened at the cursor, partly obscuring the text under it. When the return key was pressed, the window disappeared and the corrected text was visible.

The system also allowed for logging of the subjects' keystrokes as well as the times taken between each keystroke.

The editing procedure

The subjects were instructed about the following three methods for editing the text.

- The type over method, where new text could be typed over existing text, without using any commands.
- The delete and insert method, where incorrect text could be deleted by a 'delete' command and correct text could be inserted by an 'insert' command.
- The substitute method, where incorrect text could be substituted by typing in the correct string already when issuing the command.

Each task could in principle be solved by either using only *one* command or typing over the existing text.

Learning took place during two days, with an interruption of 2–3 days between each learning phase.

During the first day, the subjects were taught one method at a time, in the following order: type over, delete/insert, substitute. Each method was practised during half an hour before turning to next method.

During the second day, the subjects first repeated the methods already learnt by performing two tasks for each method. After this repetition, they were instructed to choose the best method for each coming task.

Tasks

During the first day, the tasks were given in the following order:

- type-over tasks (where the subjects were instructed to type over),
- insert/delete tasks (where the subjects were instructed about the insert/delete commands),
- substitute tasks (where the subjects were instructed about the substitute command).

During the second day, different tasks were given for which different methods would be the most efficient. First, two rehearsal tasks of each type were presented.

Then the tasks were presented as follows: Tasks which favoured the insert/delete method were spread over the whole session. Interspersed with these tasks were tasks, which could be solved more efficiently by the other two methods (type over and substitute). These alternative tasks were introduced in a systematic way, so that the subjects should have a chance detecting the useability of the alternative method. The number of consecutive tasks where another method was more efficient was systematically increased. Only two methods were mixed at a time. The delete/insert method was always one of them. The actual procedure is presented in table 3.

Subjects

As subjects served 25 psychology students, who participated in the study as part of their course requirement. Most of them got an additional amount of money, since the study was rather time-consuming.

Table 3
Order of tasks during the second day (after the rehearsal tasks).

Tasks	Method	Tasks	Method
1–18	insert–delete		
19	type	87	substitute (repeated
20–24	insert–delete		in 91 and 93–96)
25–26	type	88–90	insert–delete
27–30	insert–delete	91	substitute
31–33	type	92	delete
34–38	insert–delete	93–96	substitute
39–43	type	97	insert
44–49	insert–delete	98–99	substitute (repeated in
50–54	type		the following)
55–69	insert–delete	100	insert
70	substitute (repeated in 73)	101	substitute
71–72	insert–delete	102	insert
73	substitute	103	substitute
74–75	insert–delete	104	insert
		105	substitute
76–85	substitute (three	106	insert
	different ones	107	substitute
	repeated)		
86	insert	Note that the substitute tasks had to be	
		repeated in order to be meaningful.	

Other measurements

The subjects also underwent the following measurements:
- Raven's progressive matrices (to control for general reasoning capacity),
- Eysenck's personality inventory (introversion/extraversion and neuroticism),
- Short-term memory tests (to control for concentration and STM).

These measurements were performed before administrating the wordprocessing tasks (except the STM tests, which were performed after as well). The results of these measurements will not be reported in this article.

Further, the subjects were asked to rate themselves at the beginning of the first day and at the beginning of the second day. The results of these scales will not be treated here either.

Results

Range of latencies

The combined results for the time measurements for two different commands over all tasks and both days will first be presented. These results can be compared to the time taken for the mental operator suggested by Card et al.

Table 4
Time taken for the different steps during different commands. Data averaged over both learning days.

	Delete	Insert
Identify task	3.10	3.16
Plan method	2.52	2.83
Select command	1.28	1.15
Select argument	0.68	1.12
Close command	0.56	0.84
Evaluate result	2.44	2.49

In order to comply to the GOMS model, all trials in which the subjects made errors or issued an 'undo'-command were eliminated. Table 4 gives an overview over the average time taken for the different observed actions and the two different commands chosen to be focused, i.e. delete and insert. The average is based on 40 tasks for the delete command and 50 for the insert command. Since 25 subjects are involved, this makes 1,000 tasks in all for the delete command and 1,250 tasks in all for the insert command.

When comparing these figures to those found by Card et al., it should be noted that the measurements are averaged over a learning period (learning will be analyzed later). Further it should be noted that the learning here is rather short, and that the subjects cannot be regarded to be very skilled in their performance of the tasks. Thus the figures found here should be higher than the figures reported by Card et al.

The figures for some of the observed actions are indeed higher than the estimate for the mental operator calculated by Card et al. (1983: ch. 8) to 1.35. If the time taken for typing (roughly equivalent to 'select argument', i.e. 0.68 and 1.12 for delete and insert respectively) is subtracted from the observed times, the actions which take longer than Card et al.'s mental operator are those required for identifying the task and for planning the method. The time taken for selecting a command is surprisingly small, suggesting that the command has already been chosen, perhaps simultaneously with planning the method, as was also suggested in the task analysis above. The time for closing the command, at least for the delete command, is in the range of the time calculated by Card et al. for the physical keystroke, i.e. 0.60. Any mental operator that is involved must thus go on in parallel with issuing this keystroke.

Learning

The microstructure of learning will now be reported. The data for this analysis was calculated in the following way. Only tasks in which the subjects used the simple commands 'delete' or 'insert' were considered, according to the method-focusing decision presented above. In order to get more stable data, averages over five such tasks were calculated as a 'block' of tasks. This was made because time measurements have a tendency to be positively skewed. This is due to some single long latencies which

Table 5
Characteristics of learning, fitted to a power function.

Command	Action	Multi plicative constant	Power constant	Correlation
Delete	Total	13.56	0.91	−0.91
	Identify task	4.34	0.92	−0.76
	Plan method	3.74	0.91	−0.89
	Select command	2.14	0.88	−0.91
	Select argument	2.12	0.90	−0.82
	Evaluate command	2.16	0.90	−0.84
	Evaluate result	3.16	0.94	−0.94
Insert	Total	14.62	0.96	−0.94
	Identify task	3.02	1.01	0.17
	Plan method	3.66	0.95	−0.95
	Select command	1.66	0.93	−0.87
	Select argument	1.67	0.93	−0.90
	Evaluate command	1.32	0.92	−0.94
	Evaluate result	3.30	0.95	−0.94

might be due to occasional inattention (or system error). In order not to get an inordinately great effect from these deviating time measures, the blocks were formed. (It should be noted that other tasks might also have been performed within the block, not using the 'delete' or 'insert' commands.) Subjects could of course utilize other methods to solve their tasks, even though they were instructed to use a particular method. Thus the subjects did not perform the same number of tasks with the help of the 'delete' or 'insert' commands. For the 'delete' command, all subjects produced at least eight blocks. For the 'insert' command, at least ten blocks were produced. Three of these blocks belonged to the first day for both command types, whereas five blocks for the 'delete' command and seven blocks for the 'insert' command pertained to the second day.

It was found that the two days could be combined for the purpose of calculating the learning curves. A fit was calculated, using a linear, a power and a logarithmic equation. The data are not fine-grained enough to really discriminate between different monotonic nonlinear functions. However, the fit was marginally better in most cases with the power function. This is also the kind of learning curve suggested by Anderson (1982) as well as Card et al. (1983). The power function will therefore be chosen for table 5, where the constants and correlations for the different commands and different actions are given.

First, it should be noted that all curve fits are significant, except those for the action 'identify task', which did not follow any of the attempted curves. This action did not show any consistent learning effect, particularly not for the 'insert' command. The curves for the other actions follow each other remarkably well. It is interesting to note the similarity in both the multiplicative and the power constants between the delete

Table 6
Estimated times for basic operations. All data combined.

	Read	Plan	Recall	Constant	Total	
					Est.	Observed
Identify (del)	1.68	0.95		0.62	3.25	3.10
task (insert)	1.26	1.18		0.98	3.58	3.16
Select (del)		0.95	0.81	0.62	2.38	2.52
method (insert)		1.18	0.42	0.98	2.58	2.83
Select (del)			0.81	0.62	1.43	1.28
command (insert)			0.42	0.98	1.40	1.15
Select (del)				0.62	0.62	0.68
argument (insert)				0.98	0.98	1.12
Close (del)				0.62	0.62	0.56
command (insert)				0.98	0.98	0.84
Evaluate (del)	1.68			0.62	2.30	2.44
result (insert)	1.26			0.98	2.24	2.49

and insert commands. For the different subactions the corresponding power constants are very close to each other, although the actions start from very different points. This means that the learning curves are very similar for the different actions, although the learning amount differs somewhat. Since the observations of the actions within each command are derived from the same tasks, there is no reason to expect the learning curves to be different. However, a difference between the commands might be expected.

More basic operations

The calculations performed so far have all been based on observed data. The model of more basic operations, presented in table 2 above, will now be tested. To start with, the assumptions were first tested against the combined data, i.e. the latencies for each of the commands, combined over subjects, blocks and days. A multiple regression analysis used to estimate the values of the suggested basic operations gives the results shown in table 6.

The squared multiple regression coefficient is 0.98 ($p = 0.02$) for the delete command and 0.97 ($p = 0.08$) for the insert command. Even though the data represent a mix of subjects and learning, the model thus gets some support, although the support is weak for the insert command.

The estimated mental times are smaller than the times estimated by Card et al., except for the reading operation. The reason for this is of course that the model aims at arriving at some more basic elements of mental operators than was aimed at by Card et al.

More homogeneous sets of data might give different fits to the model. The observed times from the different days are given in table 7.

Table 7
Time taken for different actions during different commands and days.

	First day		Second day	
	Delete	Insert	Delete	Insert
Identify task	3.47	2.92	2.88	3.26
Plan method	3.09	3.43	2.28	2.57
Select command	1.67	1.45	1.06	1.02
Select argument	0.80	1.48	0.61	0.96
Close command	0.72	1.13	0.47	0.72
Evaluate result	2.69	3.03	2.30	2.26

A multiple regression fit to these data gives the estimates of the proposed operations reproduced in table 8.

The fit of four parameters to six points does not allow for much variation. However, the fact that the fit of the model lies within the range of 10% probability of getting such a fit by chance for three out of four of the more homogenous sets of data gives some support to the model. We shall see examples in the still more homogeneous sets of individual data (Individual data section) that the model can indeed be refuted.

The estimated times taken for the different operations will now be studied somewhat closer. First, the longest time is taken for the first operation, here called 'read'. This operation is similar to the 'acquire task' operation in Card et al.'s model, but the estimate here is a little lower than theirs, which amounts to 1.92. This is to be expected, since the whole action of acquiring a task in the model proposed here also includes some planning.

Further, the estimated times for most operations decrease from the first to the second day. This includes the 'read' and 'plan' operators, which are included in the action 'identify task'. In the learning data (table 5) it was found that the performance times for the full action did not decrease significantly. This suggests that the action 'identify task' is more complex than suggested here and involves operations which might not be subject to learning.

Table 8
Estimations of time for operations and model fit for day 1 and day 2 separately.

	Read	Plan	Recall	Constant	R^2	p
Delete						
Day 1	1.79	0.82	0.87	0.50	0.98	0.05
Day 2	1.30	0.43	0.60	0.42	0.97	0.08
Insert						
Day 1	1.29	1.00	0.53	1.10	0.90	0.28
Day 2	1.01	1.09	0.29	0.80	0.98	0.07

Table 9
Number of subjects where the proposed model gave a significant fit, totally and for each operation.

	Read	Plan	Recall	Constant	Total	Number of subjects using method
Delete						
Day 1	16	12	8	13	20	25
Day 2	17	12	7	10	21	24
Insert						
Day 1	3	7	0	8	9	21
Day 2	13	18	3	16	20	25

The 'plan' and 'recall' operations behave somewhat randomly with respect to learning. There are at least two possible reasons for this. Firstly, the second day required a new action to be performed, i.e. the selection of a method. This might affect the 'plan' operation. Secondly, it is probable that the 'plan' and the 'recall' operations are performed in parallel, in particular after some learning has taken place. There will be further evidence for this proposal later, in the individual data.

Individual data

Since the data reported above represent an average over all subjects, it would be interesting to test each individual subject separately. There is no reason to assume that all subjects use the same basic operations according to a single model. The model was tested for each individual subject for each command and each day separately, in the same way as for the average given in table 8 above. Table 9 gives the number of subjects who have a significant fit ($p < 0.10$) to the model, as well as the number of subjects, for whom the proposed basic operations contributed significantly ($p < 0.10$) to the regression.

We see that the model as a whole gives a significant fit for at least 80% of the subjects using the method in three of the four cases tested. This is much more than expected by chance (10%).

The worst fit is related to the insert command and the first day, where there are still more significant fits than expected by chance. A closer look at the data suggests that the poor fit for the insert command might be due to the fact that the typing and/or the closing of the command takes longer than warranted in the model. This can be seen in the averages given in table 7, but is still more evident for the subjects, where the fit was not significant. Lack of typing skill will of course affect the insert command more than the delete command.

The operation 'recall' is the one which gives the fewest incidences of significant contribution (about chance for the insert command). The data thus suggests that the recall of the command may not require an extra operation. It may instead be performed already while planning the method, as was already suggested above from the average data.

There were three subjects, for whom the model did not fit the data in any of the four cases tested. Looking closer at their data, it was found that the latencies for their last action (evaluation of result) was longer than for their first action (identifying task). This finding suggests that these subjects deviated from the model because they acted first and reflected then. The model proposes that people reflect first and act then. It was thus refuted by these subjects.

Discussion

The microstructure of people's learning and performance related to a wordprocessing system has here been approached. On the one hand, direct observations of part actions during an editing task have been studied, on the other, a model has proposed some more basic operations lying behind the actions.

The learning curves fitted to the directly observed part actions indicated that all these except one follow a similar course. The learning curves were most often best fitted by a power function, a result which complies with other studies of learning, using latencies as dependent variable. The similarity of the power constants of the curves pertaining to each command can be expected if the part actions are related to each other, and only represent an arbitrary way of dividing the total task. However, the similarity is not necessarily expected for the different commands.

The model proposed to lie behind the part actions represents an attempt at dividing the mental operator into more basic operations on the basis of hypotheses about the *cognitive* task performance. It should be noted that these more basic operations cannot be proposed to be 'elementary' or 'primitive' in all circumstances. It is impossible to suggest a single base for elementary operations without specifying the kind of performance modelled and the observations performed. In particular, suggesting operations lying far below the actions observed, such as for instance single nerve firings, is uninteresting for a model of cognitive performance and would not allow for testing with the kind of observations used here.

The main reason for searching 'more basic' *cognitive* operations lies in suggesting concepts which can be used for developing a theory of learning and cognitive performance in simple computer-related tasks.

This theory would then be possible to use both in a GOMS analysis and in a production rule analysis. It should be noted that the six original variables were only due to measurement opportunities, whereas the four suggested in the model are related to several hypotheses about people's cognitive actions.

Of course, it would have been desirable to obtain more data points in order to derive the theoretical parameters, which, in our study, do not have many degrees of freedom. Still it was found that it was possible to reject the model for some of the subjects.

There are some different conclusions to be drawn directly from the observed data and from the basic operations derived from the model. From the observed data it could have been suggested that the action of 'typing' took longer for the 'insert' command than for the 'delete' command, a rather trivial observation. In the model typing is reduced to a non-interesting constant.

A little more interesting was the direct observation that the closing of the command took longer for the 'insert' command than for the 'delete' command. Also this direct observation is rather trivial, since the user has to read more in order to check the inserted text than to check the deleted characters. The model does not contain any single operation related to the closing of the command.

Still more interesting was the direct observation that the selection of a command did not take much more time than direct typing of the argument to the command. In terms of the model, this finding suggests that the recall of a command is not an operation which can be separated from the planning of the method. The test of the model on the individual data suggests that the parallelism of these two operations was at hand for the majority of the subjects.

That there would be individual differences could have been predicted. The test of the individual data against the model showed that there were at least seven subjects out of twenty-five, in each of the four cases tested, who followed the model of taking some extra time for recalling the command. It is interesting to note that only three subjects did *not* follow the model in any of the tested cases. The model assumed that people plan first and act then. The deviating subjects were found to act first and reflect later. This might be an individual difference which has to be taken into account both when designing systems and when teaching system use.

To conclude, it seems fruitful to continue the search for more basic operations. The lesson learnt here is that the relation between the 'plan' and the 'recall' operation as suggested here has to be further investigated. More fine-grained observations would then be needed so that more degrees of freedom can be obtained. It would also be interesting to test the generality of the model by using several kinds of tasks.

The general hypothesis of the model, i.e. suggesting that people think first and act later, proved both to be supported by most subjects and to be refuted by some. It would be interesting to see what the characteristics are that encourage ways of working which deviate from the sequence suggested here. These characteristics can be useful to a theory of human learning, planning and action which takes individual differences into account, as well as suggestions for designing systems according to users' personal needs.

References

Anderson, J.R., 1982. Acquisition of cognitive skill. Psychological Review 89, 369–496.
Card, S.K., T.P. Moran and A. Newell, 1983. The psychology of human–computer interaction. Hillsdale, NJ: Erlbaum.
Polson, P.G. and D.E. Kieras, 1985. 'A quantitative model of the learning and performance of text editing knowledge'. In: L. Borman and B. Curtis (eds.), Human factors in computing systems, CHI '85. New York: ACM. pp. 207–212.
Robertson, S.P. and J.B. Black, 1986. Structure and development of plans in computer text editing. Human–Computer Interaction 2, 201–226.
Singley, M.K. and J.R. Anderson, 1985. The transfer of text-editing skill. International Journal of Man–Machine Studies 22, 493–423.
Singley, M.K. and J.R. Anderson, 1987–1988. A keystroke analysis of learning and transfer in text editing. Human–Computer Interaction 3, 223–274.
Wærn, Y., 1985. Learning computerized tasks as related to prior task knowledge. International Journal of Man–Machine Studies 22, 441–455.

Graphics design

Graphics design

Acta Psychologica 78 (1991) 307–325
North-Holland

The constraint satisfaction approach to design: A psychological investigation *

Françoise Darses

C.N.R.S., Paris, France and INRIA, Le Chesnay, France

More and more CAD systems are constraint-based. This raises the question whether this view is compatible with the cognitive processes involved in design. This paper presents a first psychological assessment of the constraint satisfaction approach. An empirical study, conducted in the domain of computer network design, outlines the results of the comparison between constraint-directed reasoning and designers' cognitive processes. It is stressed that some of the artificial intelligence (AI) formalisms which support constraint-based systems are relevant to the characteristics of psychological behaviour. Furthermore, the concept of constraint is not sufficient to describe the mental processes of design: some additional cognitive concepts have to be developed and taken into consideration in the implementation of CAD systems.

Introduction

When designers report their design activity, they tend to describe the problems they have to cope with in terms of constraints: the technological properties of the equipment, the priority to be given to particular characteristics of the problem, the order in which the actions have to be achieved, the specifications expressed by the customer – all these can be viewed as constraints to be satisfied.

Up to now, models of design have essentially described the cognitive processes involved in design as the result of planning activities

---

* This paper expands work first presented at ECCE5 (Fifth European Conference on Cognitive Ergonomics, held in Urbino, Italy, September 1990). The research associated with this paper is being carried out in the context of the Esprit Project (No. 2474) MMI2 'A Multi-Modal Interface for Man–Machine Interaction with knowledge-based systems'. The views expressed here are, however, those of the author. Special acknowledgements are due to P. Falzon and J.M. Hoc for helpful comments on earlier versions of the paper, and to Stephen Gibbons for his assistance in the English language version.

Author's address: F. Darses, C.N.R.S. Paris 8, U.R.A. 1297, Equipe de Psychologie Cognitive Ergonomique, 2 rue de la Liberté, 93526 Saint-Denis, France.

(where the development of a solution is viewed as the organization of top-down and bottom-up steps). However, the crucial role of constraints is always referred to in these models. In the opportunistic view of the design process, constraints are assimilated to the set of initial data, as specifications of the problem (Hayes-Roth and Hayes-Roth 1979). On the other hand, the top-down approach stresses the role of constraints in the management of the interactions between subproblems, which is known to be crucial in design (Simon 1969; Stefik 1981). In fact, the conflicts which emerge in the course of design are often expressed in terms of constraints by designers themselves. The role of constraints in relation to the problem is also highlighted in the evaluation activity developed by designers when they judge the successive states of solutions (Bonnardel 1989; Eastman 1970). Although it is not a constraint-directed system, Sacerdoti's program (1974) formalizes this evaluation knowledge by introducing criticalities. These are a set of procedures the role of which is to control the effects of any action which could jeopardize the plan. Criticalities are very close to constraints, since they are the formulation of the specifications of the problem domain, their weight being the expression of their importance in the problem-solving process.

Assuming that the notion of constraint is at the core of the design process, an increasing number of CAD systems have been developed using a constraint satisfaction approach. The assumption of this approach is that the solution process is viewed as the progressive reduction of an initially very broad search space. The problem is to find a means to achieve this reduction. A way of coping with this is to think that, as soon as a constraint linked to the problem is formulated, the object to be designed is partially described. However, these CAD systems have not yet been psychologically and ergonomically evaluated. The question is to know whether designers effectively represent to themselves the problem they have to cope with as a set of constraints to be satisfied.

In the second section of the paper a brief review of constraint-based CAD systems is presented, as well as some principles of the constraint satisfaction approach. The compatibility between the constraint-directed reasoning implemented in these systems and some of the cognitive characteristics of the design process are then assessed. This assessment has been conducted in the domain of computer network design. Therefore, many of the examples presented in this paper are taken from this design activity. In the fourth section some additional concepts to be taken into consideration for a constraint satisfaction approach of design are suggested.

CAD systems based on a constraint satisfaction approach

The rationale of the constraint satisfaction approach is that the functional specifications, the limitations of the resource and the design criteria can all be treated as constraints on the final product. A number of systems have been developed from this idea in order to model or assist design activities. The domains of application are described in the first part of this section. In the second part, the representation and the use of the concept of constraint, as it has been formalized in the constraint-based approach, is explained.

Domains of application

There are some professional situations where the problem is to find the best way of managing resources: for instance, a scheduling task or a planning task in the industrial domain. Here, experts proceed by trying to limit the set of constraints linked to the problem. Fox et al. (1982) report that experts spend approximately 80 to 90% of their time determining the constraints in the environment that will affect their scheduling decisions (use and resources required, technical specifications or economic and human considerations) and only 10 to 20% of their time actually constructing and modifying the schedules. Recent research has been conducted in the domains of scheduling factory jobs (Chandra and Marks 1986; Fox et al. 1982), manufacturing plans for mechanical parts (Descotte and Latombe 1985), designing experiment plans (Oplobedu et al. 1989) or planning peptide synthesis (Janssen et al. 1989). These studies have attempted to handle the problem-solving process by representing and managing the constraints involved in design.

Spatial arrangement problems are another example of a situation where some specific resources have to be properly allocated. In these problems, the number of objects and/or the number of possible locations per object is too large to permit direct solution. From this point of view, in the domain of architecture, Alexander (1972, cited in

Quintrand 1985) suggested considering the design problem in terms of requirements and elementary needs structured through the analysis of their mutual relationships. The design problem-solving process is seen as the description and management of the constraints linked to the problem. This constraint-based approach to architecture design has inspired many applications, for instance PROMET (Maroy 1971, cited in Quintrand 1985). This program provides a breakdown of the constraints through a tree representation. The subsets looked for are determined by the lowest branches of the tree, while the highest allow for the definition of three or four principal systems of constraints. A more recent study was developed by Manago (1985) in a system named LEGO. The task of this system is to complete and modify the original sketch of a building so that all the constraints are respected. A number of other space planning problems have been formalized and implemented using a constraint-directed approach, such as the determination of protein structures (Brinkley et al. 1987), furniture arrangement (Berlandier 1988) or bathroom design (Willey 1981, cited in Brinkley et al. 1987).

The constraint-based approach to the design process

Considering the process of design as the integration and the satisfaction of constraints from a number of sources involves formalizing the notion of constraint. From the constraint satisfaction paradigm, the concept of constraint can be understood as a relation between variables. Constraints can be manipulated by three different operations: constraint formulation, constraint propagation, and constraint satisfaction.

The concept of constraint as a relation between variables

The basic idea is that a design problem can be represented as a set of variables, each to be instantiated in an associated domain. For these variables there exists a set of constraints limiting the set of values allowed for specified subsets of the variables (Mackworth 1987; Nadel 1986). Consequently, *a constraint is viewed as the set of possible combinations between the values of its associated variables.*

For instance, in the domain of computer network design (Darses 1990a), the constraint *'length _compatible'* binds the variable *'type_ of_cable {thin; thick; optic fiber}'* and the variable *'segment_length {up*

Table 1
The description of the *constraint 'length _ compatible'*.

	Type _ of _ cable		
	Thin	Thick	Optic fiber
Segment _ length			
up to 185 m	×	×	×
up to 500 m		×	×
up to 3000 m			×

to 185 m; *up to 500 m*; *up to 3000 m*}'. Then, according to this relationship between the two variables, there are *six possible values of this constraint* (see table 1) which represent *six different* partial *descriptions for the network* to be cabled. [1]

This definition of the concept of constraint is consistent with a remarkable characteristic of design situations: there is no unique solution to a design problem. Different final solutions can be allowed, each being as good as the other, depending on the designers' choices. *The concept of constraint*, if it is understood as the set of the various possible combinations of the values that variables may compatibly take on, *allows for several possible solutions to the same initial problem*: since the instantiations of the constraints can differ according to the individual choices of the designers, many different solutions can be drawn.

Moreover, the definition of a constraint as a set of possible values to be progressively restricted throughout the process makes it *possible to delay the decisions to be taken on the final constraint value*. This refers to *the least-commitment strategy* often adopted by designers, which allows them to postpone their final choices.

Operations on constraints

The constraint satisfaction approach distinguishes three operations acting upon the constraints (Meseguer 1989; Stefik 1981): constraint formulation, constraint propagation, and constraint satisfaction.

Constraint formulation is the addition of a new constraint in the design process. The traditional constraint satisfaction approach works with a fixed number of constraints that are all known at the beginning.

[1] This example refers to a *binary constraint*: two variables are linked together. A special case of the definition is considered in the design process – that of *unitary constraints*. A unitary constraint is expressed by the instantiated value of a single variable, e.g. 'number_ of _ floors_ to_ cable = 3'.

However, designers' behaviour is very different: when starting to solve a problem, the set of constraints is not entirely known: the constraints are usually not fully specified at the beginning but are progressively expressed during the problem-solving process. For this reason, some authors suggest a progressive refinement of the constraints (Stefik 1981) or their progressive introduction into the design process (Janssen et al. 1989). The verbal protocols collected in computer network design show that more information and specifications are asked for throughout the entire process. In fact, beyond the difficulty of expressing constraints, a greater complexity arises from the management of constraints during the planning process.

Constraint propagation is the creation of new constraints from old constraints. When constraints are propagated, they bring together new requirements from separate parts of the problem. These modifications can be the addition or the suppression of some values triggered by the formulation of the new constraint. For instance, the propagation of the constraint *'length _compatible'* creates a new constraint which could be called 'environmental_ compatible'. For this constraint, the type of cable to be chosen (*type _of _cable {thin, thick, optic fiber}*) depends on the cable location (*cable _location {inside or outside}*). By satisfying this new constraint, a restriction of the initial set of possible values is obtained.

Constraint satisfaction is the operation of finding the appropriate values for the variables linked together in a constrained relationship. When a constraint is satisfied, a part of the solution is described more precisely; thus, the set of possible solutions is delineated more accurately.

Different types of constraint

The necessity to account for the fact that designers rank constraints according to their importance or their priority has been introduced in computational systems in several ways. Descotte and Latombe (1985) attach weights to the constraints. Constraints are expressed as production rules: each right-hand side of the rule is weighted according to the importance of its satisfaction in the human expert's view.

Janssen (1990) has distinguished *validity constraints* from *preference constraints*. A validity constraint must be verified in the solution for the solution to be admitted. Most validity constraints concern technological prescriptions, topological prescriptions or customer specifications. For instance, the compatibility between the type of cable and the length of the segment to be cabled is a validity constraint. But, within the set of allowable solutions, designers also choose one solution rather than another using preference constraints. Whether a preference constraint is applied or not, the object to be designed remains an allowable object. For instance, in computer network design, a preference constraint may be applied in order to allow the possibility of extending.

Compatibility between constraint-directed reasoning in CAD systems and the cognitive characteristics of design

Although most of the research efforts in artificial intelligence (AI) are aimed at improving techniques and methods (backtracking algorithms, consistency algorithms, cooperative algorithms, constraint propagation, etc), CAD systems based on a constraint satisfaction approach have attempted to take into consideration some characteristics of human problem-solving processes. Some of them allow a flexible and progressive introduction of constraints in the design process idea (Janssen et al. 1989) or recommend distinguishing among the constraints according to the way that designers deal with them. But these systems have not yet been psychologically and ergonomically evaluated. The question is to determine whether designers actually view the problem which they have to solve as a set of constraints to be satisfied.

In order to make an initial assessment of the constraint-based approach to design, a pilot study was conducted in the domain of computer network design. The data collected were examined to identify more general principles for the elaboration of a theoretical model. But this empirical study highlights some interesting findings regarding the compatibility between constraint-directed reasoning, as emphasized in AI, and some cognitive characteristics of the design process (Darses 1990a, b). The main results are presented in the following sections.

The empirical study

In network design, experts have at their disposal initial data, such as the type of machines to be connected, their location, the plan of the building and some specific requirements listed by the customer. From these data, choices must be made (Darses and Falzon 1989): topological (location of the cables in the building), technological (specific properties of the network equipment), functional (architecture of the network).

The pilot study consisted of asking six designers (working in the same company) to deal with the same problem of network configuration. Valuable information about the management of constraints can be derived from the comparison of problem-solving processes with various levels of expertise. Three of the subjects were experienced designers and three of them were novices. The designers in charge of designing the computer networks in the company were the experienced subjects. People whose professional competence is related to network design–network management, software development, computer sales and management were the novices; thus, the subjects had a basic knowledge of computer network design, because the domains are very interdependent. The data collected were verbal protocols obtained through simultaneous verbalization. A methodology of analysis (Darses 1990a) based upon the constraint-based theoretical framework was applied: protocol description focused on highlighting constraint formulation, as well as constraint propagation and satisfaction. The results allow an initial comparison between the constraint-based approach, as it is modelled in current CAD systems, and some of the cognitive processes involved in design.

Constraint as a relation between variables

From a psychological point of view, the first question is to determine if it is relevant to formalize the concept of constraint as a relationship between variables as suggested in AI.

The analysis of the protocols shows that *about half of the utterances in the verbal protocols (46%) can be expressed as a relation between variables* (see table 2). [2] Thus, the formal description of the constraint provided in AI appears to be relevant to the psychological representation of the concept of constraint.

---

[2] The *constraint* units include both unitary and binary constraints expressed by designers during the design process.

Table 2
Use of the *constraints* in relation to the *other units* depending on level of expertise (%).

Categories	Subjects							
	Experts				Novices			
	A	B	C	Mean	D	E	F	Mean
Constraints	53.2	52.2	44.7	50.0	41.8	35.4	50.0	42.4
Other units	46.8	47.8	55.3	50.0	58.2	64.6	50.0	57.6

Nevertheless, in this pilot study there is a small difference between experienced and novice designers in their ability to express the constraints (50.0% against 42.4%). This suggests that novices may have greater difficulty in expressing their knowledge in terms of constraints. But because of the small number of observations this difference may be explained by interindividual variability; it may be attributed to subject E, whose results appear to be atypical in comparison with the other individual results. Further investigation will have to be made so as to understand this difference.

Operations on constraints

Beyond the processes of *expressing* the constraints linked to the problem, a question arises about how the constraints are *managed* during the process of design. From a psychological point of view, it is a matter of assessing how relevant it is to apply to constraints a set of operations – formulation, propagation and satisfaction, as suggested in AI formalism. Since formulation, propagation and satisfaction of constraints can be broadly characterized as inferences in the problem-solving process, the different roles these operations play have been interpreted in the empirical investigation. Once again, we must take these data as a first illustration of our theoretical model; in the results below, interindividual differences within each group mean that generalization is not yet possible.

The analysis of the protocols has stressed *three occurrences*: constraint *only formulated*, constraint *formulated and simultaneously satisfied*, constraint *formulated then satisfied* (the satisfaction is delayed and occurs later on in the process).

Table 3
Management of constraints depending on level of expertise (%).

	Experts				Novices			
	A	B	C	Mean	D	E	F	Mean
Formulation only	26.0	36.0	56.5	39.5	52.5	33.5	87.0	57.7
Formul./satisfaction simultaneous	53.0	28.0	12.0	31.0	34.5	66.5	6.5	35.8
Formul./satisfaction delayed	21.0	36.0	31.5	29.5	13.0	0.0	6.5	6.5

Moreover, it has been highlighted that *these methods of managing the constraints differ greatly according to the level of expertise of the designers*.

As it is shown in table 3, *the occurrence 'formulation only' is readily used by the novices* (57.7% against 39.5%). This result [3] confirms that the difficulty of dealing with constraints is not related to their formulation; the novices are able to express constraints as the experienced designers do. But the difficulty lies in constraint management: the novice designers seem to meet some problems as soon as propagation and satisfaction have to be applied to constraints.

The occurrence 'formulated and simultaneously satisfied' appears in a *comparable proportion regardless of level of expertise* (31.0% against 35.8%). But a closer examination of the protocols shows that the constraints which are managed in this way *differ by their nature*, depending on the level of expertise of the designers. This point will be developed in the next section.

On the other hand, there is an *important difference* between the designers *for the occurrence 'formulation and satisfaction delayed'* (29.5% against 6.5%). It is not surprising that novices are less able to delay the satisfaction of constraints. The ability to postpone the satisfaction of constraints is more a characteristic of experienced designers. They try to keep open as long as possible the choices to be made in the design process and try to delay decisions, so as to avoid premature and inappropriate decisions which would narrow the possibilities and create unsolvable conflicts between subprob-

---

[3] These results (and those reported in tables 4, 5 and 6) refer to the set of binary constraints expressed in the verbal protocols.

lems. It has been observed that novices tend to compromise themselves early in the solution process and do not adopt this strategy,
known as the least commitment strategy (Lebahar 1983; Sacerdoti
1977). Actually, our results illustrate that novices find some difficulties in developing such a strategy.

Types of constraints

Previous psychological work has shown that constraints can be
specified according to the way they are involved in the design process.
Eastman (1970) stressed a distinction between constraints given in the
problem statement, constraint retrieved from the documents and
constraints recalled from memory. Analyzing the activity of architectural designers, the author noticed that their competence depends on
the type of constraint they deal with and on the way they have
organized and structured these constraints in memory. Similar distinctions were also adopted by Bonnardel (1989) in the domain of
aerospatial structures design; the author points out other types of
constraints: those which can be deduced from previous constraints or
from the current state of the solution.

Our study suggests considering these distinctions from the point of
view of the propagation process. While the prescribed constraints can
be formalized as unitary constraints the value of which has been
previously set (e.g. 'the building is six floors high' or 'we want optic
fiber to be used'), the deduced constraints must be built into binary
relationships and have their instantiation submitted to the decision of
designers (for example, it is up to the designer to choose the type of
cable according to the length of the segment to be cabled).

AI research has also suggested distinguishing more accurately the
type of constraint handled in the course of the design process. They
have focused on the necessity of distinguished constraints according to
their importance and their priority when solving the problem. As
stated above, this has been formalized in CAD systems either by
attaching weights to the constraints (Descotte and Latombe 1985), or
by defining specific types of constraint, e.g. preference constraints and
validity constraints (Janssen 1990).

The psychological relevance of distinguishing constraints on the
basis of their preference and validity nature has not yet been

assessed. Nevertheless, we observed that the development of the solution is closely connected to designers' own choices and preference criteria. Therefore, we found that our empirical investigation would provide a good framework for evaluating how relevant it is to make a distinction between preference constraints and validity constraints.

A first remark is that the preference or validity nature of constraints depends on the specific characteristics of the problem to be solved. In fact, some constraints can be considered either on validity or preference criteria, according to the case. For instance, the final cost of the network is considered as a validity constraint when the customer initially specifies a maximum cost figure. But cost can be viewed as a preference constraint if designers impose upon themselves a 'reasonable price' criterion though not specified by the customer. This characterization of the constraints may also change during the course of the design process. For the problem that the subjects had to solve, we made sure that the definition of the preference and validity constraints was clear.

Within the scope of this pilot study, our results support the psychological relevance of *distinguishing preference constraints from validity constraints*. First, the way constraints are *managed differs greatly, according to their type*. Second, an important *difference* is pointed out in the way of *handling each type of constraint, related to designers' level of expertise*.

Table 4 shows the global use of validity and preference constraints depending on the level of expertise. *Novice designers tend to use validity constraints more than preference constraints*. This result is not surprising, since we know that preference constraints are progressively built up throughout the professional experience and thus are more difficult to bring into play. Two of the experienced

Table 4
Global use of validity and preference constraints depending on level of expertise (%).

	Experts				Novices			
	A	B	C	Mean	D	E	F	Mean
Validity constraints	74.0	42.5	31.0	49.2	56.5	66.5	47.0	56.5
Preference constraints	26.0	57.5	69.0	50.8	43.5	33.5	53.0	43.5

Table 5
Validity constraints posting depending on level of expertise (%).

	Experts				Novices			
	A	B	C	Mean	D	E	F	Mean
Formulation only	6.7	33.0	60.3	33.3	46.4	50.0	86.2	60.9
Formul./satisfaction simultaneous	64.8	67.0	39.7	57.2	53.6	50.0	13.8	39.1
Formul./satisfaction delayed	28.5	0.0	0.0	9.5	0.0	0.0	0.0	0.0

designers tended to use preference constraints more than validity constraints; but the results of subject A do not enable further interpretation of the data.

The global comparison of tables 5 and 6 shows a remarkable difference between experienced designers and novices in the way they manage constraints, according to their type – validity or preference. *Experienced designers clearly differentiate the management of preference constraints from the management of validity constraints.* The way they deal with these two types of constraints is detailed below. *Novices do not show any difference in the way they handle preference constraints and validity constraints:* their way of working is rather an immediate satisfaction of the formulated constraints and delaying satisfaction is almost never applied to constraints. Again, the occurrence of 'formulation only' appears in large proportion: although novices are able to *formulate* constraints, they lack sufficient knowledge to be able to use the constraints.

Table 6
Preference constraints posting depending on level of expertise (%).

	Experts				Novices			
	A	B	C	Mean	D	E	F	Mean
Formulation only	80.8	37.4	54.3	57.5	59.8	0.0	87.8	49.2
Formul./satisfaction simultaneous	19.2	0.0	0.0	6.4	10.3	100	0.0	36.8
Formul./satisfaction delayed	0.0	62.6	45.7	36.1	29.9	0.0	12.2	14.0

For expert designers, most of the validity constraints (see table 5) *are formulated and simultaneously satisfied* in the process. Although novices often 'only formulate' validity constraints, they do also often 'formulate and simultaneously satisfy' these constraints (around 40%). *An important remark is that, regardless of level of expertise, validity constraint satisfaction is almost never delayed.* We assume that the values of these constraints are instantiated as soon as they are expressed because they represent an unavoidable commitment to be made. Thus they enable designers to restrict the search space quite easily.

The results related to the preference constraints (see table 6) are not homogeneous: the interindividual variation is important. Nevertheless, trends can be sketched: except for subject E, most of the preference constraints are not satisfied simultaneously with their formulation: they are either *only formulated* or *formulated at the beginning of the problem-solving process and satisfied later.* The occurrence of 'formulation only' can be explained in many ways: preference constraints can be evoked and then given up if they do not fit with the development of the solution or with other choices; concerning the novices, another explanation is related to lack of knowledge. Moreover, a remarkable *difference between experts and novices is shown in their competence in managing preference constraints*: the novices are unable to handle these constraints by postponing their satisfaction (14.0%), while this way of working is readily used by experts (36.1%).

Additional concepts needed

As shown in the previous section, some features of constraint-directed reasoning implemented in CAD systems are compatible with the characteristics of designers' reasoning. But the empirical study has also pointed out some aspects of the cognitive processes involved in design, which must be taken into consideration in these systems. Two facets must be elaborated so as to give an account of the psychological activity of designers: a more accurate definition of the concept of constraint is needed and the concepts of action plans and schemata should be introduced.

Constraints and rules

Constraints are formalized (Sriram and Maher 1986) in CAD systems either as a relation between variables or as rules (productions), or objects (schemas or frames). For instance, the GARI system (Descotte and Delesalle 1986) implemented for generating the manufacturing plans of mechanical parts applies a rule-based approach. A constraint is expressed as follows: 'if a hole H1 opens into another hole H2, then one is recommended to manufacture H2 before H1 in order to avoid the risk of damaging the drill'.

However, a problem arises from this formalization of the concept of constraint. Consider the following utterances. Formally, they can be expressed as production rules:

(i) 'if the length of the segment to be cabled does not exceed n, then the type of cable is x';
(ii) 'if the cable is a thin cable, and if the machines are connected through a BNC connector, then the length of the cable is the length of the segment to be cabled plus the number of connections to be done multiplied by ten meters'.

The difficulty which arises here is that these two utterances do not have the same operative meaning. In fact, the knowledge described in (i) can be used for network design by triggering either the 'type of cable' variable or the 'segment to be cabled' one. The production rule is not directional and it makes sense to say:

(i') if the type of cable is x, then the length of the segment to be cabled does not exceed n'.

On the other hand, the utterance (ii) is directional. It does not make sense to try to reverse the sentence. It would loose its operative property. Actually, the knowledge contained in the reversed sentence is unlikely to be useful for designers:

(ii') 'if the length of the cable is the length of the segment to be cabled plus the number of connections to be done multiplied by ten meters, then the cable is a thin cable and the machines are connected through a BNC connector'.

This suggests that designers might handle these two kinds of cognitive entities differently. Thus, for a constraint satisfaction approach to design, it seems necessary to distinguish on the one hand the *constraints* (non-directional relationships between variables) and the *rules* (directional relationships). A CAD system based on a constraint satisfaction approach should give the designer the possibility of dealing in different ways with the expression of constraints and the expression of rules.

Other units

Since the prominent place of constrains in the reasoning of designers has been pointed out, it is tempting to see the process of design as formulating and satisfying constraints, the succession of which progressively shapes the object to be designed until the final solution is reached. But some steps of the solution process are unlikely to be interpreted in terms of constraints. The analysis of the protocols shows that the cognitive activity occurring during design cannot be specified only in terms of constraints. There are at least two additional cognitive entities which are triggered simultaneously with constraints: action plans and schemata.

Action plans

The analysis of the protocols has stressed that designers build their reasoning process from a set of action plans (Falzon and Darses 1990), which are triggered in the course of the design process. An example taken from network design is presented in fig. 1. The top-down representation of an action plan does not prevent bottom-up steps from being developed. How constraints trigger or act on the development of action plans has not yet been investigated.

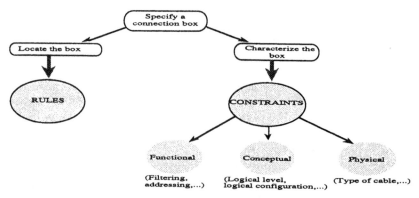

Fig. 1. An action plan for specifying a connection box.

backbone
variable 1: location of the cable variable 2: type of cable variable 3: location of junction points variable 4: logical level of the junction points variable 5: type of connection boxes variable 6: location of the connection boxes variable 7: physical configuration of the backbone

Fig. 2. The representation of the 'backbone' design object.

Schemata

Another cognitive unit to be called up in the course of the design process is formalized as a schema. We assume that designers represent to themselves a set of 'design objects'. The description of the current object to be designed is built on the basis of one of these schemata. In other words, these schemata are a kind of blue-print which provides a representation of successive states of a solution. Each of these objects has individual features, which can be formalized as variables of the associated schema. For instance in computer network design, designers handle the object 'backbone', which can be formalized as represented in fig. 2.

Constraints act on these schemata so as to give appropriate values to the variables. Constraints are also used to manage the relationship between variables belonging either to the same object or two different objects. For the object 'backbone' described above, a constraint exists which binds variable 1 and variable 2 (see the constraint environment_compatible), but there exists also a constraint which binds variable 5 of this object with another design object (not depicted here) named 'sub-network'.

Conclusion

This paper shows that many features of constraint-directed reasoning, as formalized in CAD systems, are compatible with the characteristics of the cognitive processes involved in design. The formalization of the constraint concept as a relationship between variables is consistent with its cognitive representation. It is confirmed that operations on constraints are performed by designers. A psychological distinction between types of constraints is established. On the other hand, this first investigation stresses that the constraint satisfaction approach should take into consideration additional concepts so as to improve the psychological validity of systems: unitary and binary constraints do not have a similar status in designers' minds, and relate to the notions of prescribed and deduced constraints. Besides this, some other cognitive units which figure in the design process have to be taken into consideration, such as the representation of plans of action and schemata. We assume that these enlargements would be likely to provide some ergonomic advances for the implementation of CAD systems.

References

Berlandier, P., 1988. Intégration d'outils pour l'expression et la satisfaction de contraintes dans un générateur de systèmes experts. Report No. 924, INRIA, Sophia-Antipolis.

Bonnardel, N., 1989, L'évaluation de solutions dans la résolution de problèmes de conception. Report No. 1072, INRIA, Rocquencourt.

Brinkley, J., B. Buchanan, R. Altman, B. Duncan and C. Cornelius, 1987. A heuristic refinement method for constraint satisfaction problems. Knowledge systems laboratory. Report No. 87-05, January, Computer Science Department, Standford University.

Chandra, N. and D. Marks, 1986. Intelligent use of constraints for activity scheduling. Applications of IA in Engineering.

Darses, F. and P. Falzon, 1989. The design activity in networking: General remarks and first observations. Technical report, March, INRIA, Rocquencourt.

Darses, F., 1990a. Gestion de contraintes au cours de la résolution d'un problème de conception de réseaux informatiques. Report No. 1164, INRIA, Rocquencourt.

Darses, F., 1990b. 'Constraints in design: Towards a methodology of psychological analysis based on AI formalisms'. In: D. Diaper, D. Gilmore, G. Cockton and B. Shackel (eds.), Human-computer interaction INTERACT '90. Amsterdam: North-Holland. pp. 135–139.

Descotte, Y. and H. Delesalle, 1986. Une architecture de système expert pour la planification d'activité. Proceedings of Sixth International Workshop Expert Systems and their Applications, Avignon, April 28–30. pp. 903–916.

Descotte, Y. and J.C. Latombe, 1985. Making compromises among antagonist constraints in a planner. Artificial Intelligence 27, 183–217.

Eastman, C., 1970. 'On the analysis of intuitive design processes'. In: G. Moore (ed.), Emerging methods in environmental design and planning. Proceedings of the design methods group. First International Conference. Cambridge, MA: MIT Press.

Falzon, P. and F. Darses, 1990. 'Conception de réseaux informatiques: une approche psychologique'. In: Sur la modélisation des processus de conception créative. Actes de 01 Design, Première table ronde nationale sur les nouvelles méthodes et paradigmes de modélisation de processus de conception créative. Cabourg, October 21–23.

Fox, M.S., B. Allen and G. Strohm, 1982. Job-shop scheduling: An investigation in constraint directed reasoning. Proceedings of AAAI-82, Pittsburgh, PA.

Hayes-Roth, B. and F. Hayes-Roth, 1979. A cognitive model of planning. Cognitive Science, 3, 275–310.

Janssen, P., 1990. Aide à la conception: une approche basée sur la satisfaction de contraintes. Thèse de Doctorat, Spécialité Informatique, Université de Montpellier II.

Janssen, P., P. Jégou, B. Nouguier and M.C. Vilarem, 1989. Problèmes de conception: une approche basée sur la satisfaction de contraintes. Proceedings of the Ninth International Workshop on Expert Systems and their Applications, May 29–June 2. pp. 71–84.

Lebahar, J.C., 1983. Le dessin d'architecte. Roquevaire: Parenthèses.

Mackworth, A.K., 1987. 'Constraint satisfaction'. In: S. Shapiro (ed.), Encyclopedia of artificial intelligence. New York: Wiley. pp. 205–211.

Manago, C., 1985. Lego: un système qui traite des contraintes. Proceedings of Cognitiva 85. Application à la CAO dans le bâtiment. Paris, June 4–7.

Meseguer, P., 1989. Constraint satisfaction problems: An overview. AICOM 2, 1.

Nadel, B.A., 1986. The general consistent labeling or constraint satisfaction problem. Technical report, Department of Computer Science, University of Michigan, MI.

Oplobedu, A., J. Marcovitch and Y. Tourbier, 1989. CHARME: un language industriel de programmation par contraintes, illustré par une application chez Renault. Proceedings of the Ninth International Workshop on Expert Systems and their Applications, May 29–June 2. pp. 55–70.

Quintrand, P., 1985. La conception assistée par ordinateur en architecture. Paris: Hermès.

Sacerdoti, E.D., 1974. Planning in a hierarchy of abstraction spaces. Artificial Intelligence 5, 115–135.

Sacerdoti, E.D., 1977. A structure for plans and behavior. New York: Elsevier.

Simon, H.A., 1969. 'The science of design and the architecture of complexity'. In: Sciences of the artificial. Cambridge, MA: MIT Press.

Sriram, D. and M.L. Maher, 1986. The representation and use of constraints in structural design. Applications of AI in Engineering.

Stefik, M., 1981. Planning with constraints (MOLGEN: Part 1). Artificial Intelligence 16, 111–140.

Acta Psychologica 78 (1991) 327–332
North-Holland

Author index

Acta Psychologica 78 (1991) 333–337
North-Holland

Subject index

Printed and bound by CPI Group (UK) Ltd, Croydon, CR0 4YY

03/10/2024

01040330-0008